RESEARCHES

Volume Four

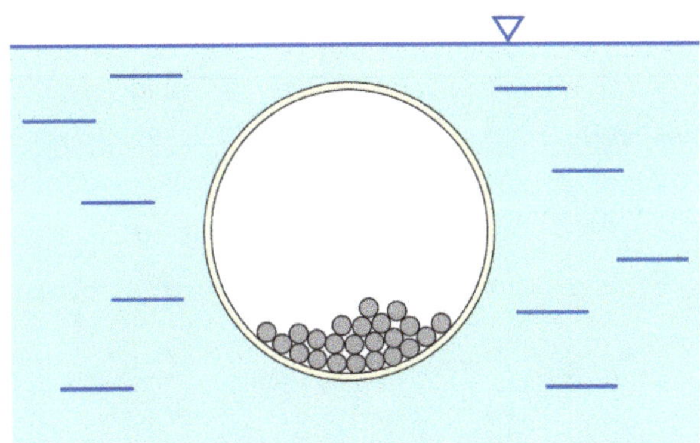

James R Warren

Written and Published in the United Kingdom by Midland Tutorial Productions

First Edition: 15 October 2022

File Prefix Code: R4

ISBN 978 1 915750 03 7

Printed and Bound by IngramSpark

Midland Tutorial Productions Publishers
31 Victoria Avenue
Bloxwich
Walsall
WS3 3HS
United Kingdom

RESEARCHES

Volume Four

First Edition

James R Warren

MIDLAND TUTORIAL PRODUCTIONS
BLOXWICH

RESEARCHES
WARREN

To The Glory of The Loving God

Who Made Our Minds Free

TABLE OF CONTENTS

PREFACE

The first chapter of Volume Four of my "Researches" returns to one of my most gratifying and arguably useful themes: That of high-accuracy approximation, preferably in closed-form, of otherwise expensive or difficult functions.

In this case, we establish a way of approximating the Complete Elliptic Integral of the First Kind (a CEIF), a function that has a number of applications in the sciences and in statistics. The method involves the application of a three-pass Arithmetic-Geometric Mean (AGM) to bounding analytic expressions.

In some contexts the provision of an accurate approximation can greatly economise the computation of a more refined solution.

The second chapter studies the feasibility of using neutrally-buoyant floats to achieve simple spot estimations of rivulet velocity. Several variations of ballast materials and techniques are calculated-out, and it is established that the problem is surprisingly delicate and certain forms of ballast provision infeasible. Like all gadgeteering, float employment is very vulnerable to the progress of technology, and I suppose that today, even in poor communities, ultrasonic techniques would more readily be applied to the assessment of stream power. Because of the several variations on the theme there is significant repetition and once again I can only apologise to readers who find such tedious or wasteful.

The third chapter is about the analytic geometry, and especially the trigonometry, of the Thalean triangle (the right-angled triangle in a semi-circle) as derived from selected central and circumferential points. Areas are checked in several ways, and some interesting trigonometrical identities emerge.

In the final chapter I take a critical look at a couple of the experimental writings of the great French naturalist Georges-Louis Leclerc, First Comte de Buffon (1707-1788). The two quantitative studies I found in my contemporary translation trespassed the physical sciences, and thus became immediate prey to this author. Both studies, the first about the attenuation of light through glass, and the second organised by his friend Tillet concerning the density of materials throw an interesting sidelight (so to say) on the science of the time and the often strange choices made as regards procedure and computation.

Enjoy.

James R Warren
Bloxwich
25 September 2022

CHAPTER ONE

The Elliptic Integral

RESEARCHES
WARREN

A Simple but Accurate Approximator of
The Complete Elliptic Integral of the First Kind

by
James R Warren BSc MSc PhD PGCE

PART I
METHODOLOGICAL REVIEW

The Complete Elliptic Integral of the First Kind, K(k), is defined in Abramowitz and Stegun[1] as:-

$$K(k) = CEIF(k) = \int_0^{\frac{\pi}{2}} \frac{1}{\sqrt{1 - x.\sin(\theta)^2}}.d\theta = \int_0^1 \frac{1}{\sqrt{(1-t^2).(1-x.t^2)}}.dt$$

Equation One

There is no perfect closed-form expression for the exact CEIF(k).

There are numerous schemes of numerical integration and many closed-form approximators that are more or less accurate estimators of The Complete Elliptic Integral of the First Kind.

There are also "general" numerical approaches such as Rule 18 that notably fail to give usable values.

Standard Approximators

Double Factorial Quotient Series

This may be summarised as:-

$$K_{series}(k) = \frac{1}{2}\pi\left[1 - \sum_{i=1}^{n} k^i \left(\frac{\prod_{j=1}^{i} 2j-1}{\prod_{j=1}^{i} 2j}\right)^2\right]$$

Equation Two

When n = 15 and k = ½ the accuracy is to five mantissa digits.

Simpson's Rule

Allow that:-

$$g_0 = \frac{1}{\sqrt{1 - k.\sin(0)^2}} = 1$$

Equation Three a

$$g_n = \frac{1}{\sqrt{1 - k.\sin\left(\frac{\pi}{2}\right)^2}} = 1.41421356237 \equiv \sqrt{2}$$

Equation Three b

$$g_i = \frac{1}{\sqrt{1 - k.\sin\left(\frac{i}{n}.\frac{\pi}{2}\right)^2}}$$

Equation Three c

Then:-

$$K_{simpson}(k) \approx \frac{h}{3}\left[g_0 + g_n + \sum_{i=1}^{n-1}\left[3 - (-1)^i\right].g_i\right]$$

Equation Four

where:-

$$h = \frac{\pi}{2n}$$

Equation Five

Simpson's Rule[2] works on the basis that there are an odd number of data values (i.e. an even number of intervals). Simpson's Rule is one of the simplest and most-elegantly programmable, as well as one of the most robust, of the family of Closed Newton-Cotes numerical integration formulae.

It is a classical scheme good for many continuous functions whose trajectories can realistically be modelled as a succession of parabolas.

When n = 48 (giving 49 data inclusive of the zeroth) and k = ½, the accuracy of $K_{simpson}(k)$ is nine mantissa digits.

The Hastings Polynomial Approximation

Hastings[3] provides a high-precision double polynomial for estimating values of CEIF(k):-

$$K_{hastings}(k) = \sum_{i=0}^{4} a_i . k^i + (\log_n k) \sum_{i=0}^{4} b_i . k^i$$

Equation Six

Table One presents the requisite coefficients of the relevant Hastings Polynomials:-

a_0	1.38629436112	b_0	0.50000000000
a_1	0.09666344259	b_1	0.12498593597
a_2	0.03590092383	b_2	0.06880248576
a_3	0.03742563713	b_3	0.03328355346
a_4	0.01451196212	b_4	0.00441787012

Table One
Hastings Polynomial Coefficients for CEIF Estimation

For k = ½ the Hastings Polynomial system yields nine-digit accuracy.

PART II
APPROXIMATIONS DEVELOPED FROM
THE ARITHMETIC-GEOMETRIC MEAN

A definition of The Complete Elliptic Integral of the First Kind in terms of The Arithmetic-Geometric Mean is furnished by the Gauss-Lagrange Formula[4]:-

$$K(k) = CEIF(k) = \int_0^{\frac{\pi}{2}} \frac{1}{\sqrt{1 - x.\sin(\alpha)^2}}.d\theta = \frac{\pi}{2.AGM\left(1, \sin(\alpha)\right)}$$

Equation Seven

which may conveniently be tabulated as:-

$$K(k) = CEIF(k) = \int_0^1 \frac{1}{\sqrt{\left(1 - t^2\right)\left(1 - k.t^2\right)}}.dt = \frac{\pi}{2.AGM\left(1, \sqrt{k}\right)}$$

Equation Eight

where:-

$$\alpha = \sin^{-1}\left(\sqrt{k}\right)$$

Equation Nine

In the context of Equation Eight, K is the Complete Elliptic Integral of the First Kind for Argument k where $0 < k \leq 1$, and we shall develop AGM processes on this interval.

π is the Ludolphine Constant and AGM(x,y) is the Arithmetic-Geometric Mean of x and y defined by:-

$$\begin{pmatrix} a_0 \\ b_0 \end{pmatrix} = \begin{pmatrix} x \\ y \end{pmatrix} \qquad \begin{pmatrix} a_{i+1} \\ b_{i+1} \end{pmatrix} = \begin{pmatrix} \dfrac{a_i + b_i}{2} \\ \sqrt{a_i b_i} \end{pmatrix} \qquad AGM(x,y) = \begin{pmatrix} \dfrac{x + y}{2} \\ \sqrt{xy} \end{pmatrix}$$

Equation Ten a   **Equation Ten b**   **Equation Ten c**

whilst:-

$$c_i = \frac{a_i + b_i}{2}$$

Equation Ten d

This iteration will yield fifteen-figure accuracy in four passes when presented with the initial conditions:-

$$a_0 = 1 \qquad\qquad b_0 = \sin(\alpha) \qquad\qquad c_0 = \cos(\alpha)$$

Equation Eleven a   **Equation Eleven b**   **Equation Eleven c**

Approximation of the AGM

A three-pass AGM can achieve a precision sufficiently high for many purposes and I dare say for all environmental science applications.

By internal substitutions it is possible to express a three-pass AGM by the construct:-

$$AGM_{(2)} = \dfrac{\dfrac{\dfrac{\frac{x+y}{2}+\sqrt{xy}}{2}+\sqrt{\frac{\frac{x+y}{2}\cdot\sqrt{xy}}{2}}}{2}+\sqrt{\dfrac{\frac{x+y}{2}+\sqrt{xy}}{2}+\sqrt{\frac{x+y}{2}\cdot\sqrt{xy}}}}{2}$$

Equation Twelve

A degree of simplification gives:-

$$AGM_{(2)} = \frac{x}{2^4}+\frac{y}{2^4}+\frac{1}{2^3}\left(x^{\frac12}y^{\frac12}\right)+\frac{1}{2^{\frac53}}\sqrt{x+y}\cdot x^{\frac14}y^{\frac14}+\frac{1}{2^{\frac94}}\sqrt{x+y+2\sqrt{x}\sqrt{y}}\cdot(x+y)^{\frac14}\cdot x^{\frac18}y^{\frac18}$$

Equation Thirteen

and further simplifications bring us to:-

$$AGM_{(2)} = 2^{-\frac54}\sqrt{x+y+2\sqrt{xy}}\cdot(x+y)^{\frac14}\cdot(xy)^{\frac18}$$

Equation Fourteen

For a system where $a_0 = 1$ and $b_0 = \sin(\alpha)$ we may note the identities:-

$$AGM_{(2)} = \frac{\left(x+y+2\sqrt{xy}\right)^{\frac12}(x+y)^{\frac14}\cdot(xy)^{\frac18}}{\left(\sqrt{32}\right)^{\frac12}}$$

$$= \frac{\left(1+\sin(\alpha)+2\sqrt{\sin(\alpha)}\right)^{\frac12}(1+\sin(\alpha))^{\frac14}\cdot(\sin(\alpha))^{\frac18}}{\left(\sqrt{32}\right)^{\frac12}}$$

$$= 2^{-\frac54}(1+\sin(\alpha))\sqrt{\sqrt{1+\sin(\alpha)}\cdot\sqrt[4]{\sin(\alpha)}}$$

Equation Fifteen

Any of these identities reduce to:-

$$AGM_{(2)} = \frac{(1+\sin(\alpha))^{\frac{1}{4}}.\left[(\sin(\alpha))^{\frac{1}{8}} + (\sin(\alpha))^{\frac{5}{8}}\right]}{\sqrt[4]{32}}$$

Equation Sixteen

The Specific Defect, $SpDef_{AGM\text{-}PA}$, is given by:-

$$SpDef_{AGM-Eqn.16} = 100\frac{Eqn.16 - AGM}{AGM}$$

Equation Seventeen

For the case of:-

$$\sin(\alpha) = \sqrt{\frac{1}{2}} = 0.707106781186548 \equiv \cos(\alpha)$$

the Specific Defect vis-a-vis the full AGM is -0.000000000048646, or 12-figure accuracy for the pseudoanalytic form.

<u>Logarithmic Development of the Approximation</u>

As adumbrated above, the resolution of Equation Twelve in terms of logarithms yields the simplification:-

$$AGM_{(2)} = \frac{\left(x + y + 2\sqrt{xy}\right)^{\frac{1}{2}}(x+y)^{\frac{1}{4}}.(xy)^{\frac{1}{8}}}{\sqrt[4]{32}}$$

Equation Fifteen a

Appropriate substitutions of x = 1 and y = sin(α) then give:-

$$AGM_{(2)} = \frac{\left(1 + \sin(\alpha) + 2\sqrt{\sin(\alpha)}\right)^{\frac{1}{2}}(1+\sin(\alpha))^{\frac{1}{4}}.(\sin(\alpha))^{\frac{1}{8}}}{\sqrt[4]{32}}$$

Equation Fifteen b

whilst further simplification of the above expression brings us:-

$$AGM_{(2)} = 2^{-\frac{5}{4}}\left(1 + \sin(\alpha) + 2\sqrt{\sin(\alpha)}\right)^{\frac{1}{2}}(1+\sin(\alpha))^{\frac{1}{4}}.(\sin(\alpha))^{\frac{1}{8}}$$

Equation Fifteen c

Continued simplification of the above expression brings us to:-

$$AGM_{(2)} = 2^{-\frac{5}{4}}\left[\left(1+\sqrt{\sin(\alpha)}\right).\left(1+\sin(\alpha)\right)^{\frac{1}{4}}.\left(\sin(\alpha)\right)^{\frac{1}{8}}\right]$$

Equation Sixteen

Whereupon substitution of $\sqrt{k}$ for $\sin(\alpha)$ gives:-

$$AGM_{(2)} = 2^{-\frac{5}{4}}.\left(1+k^{\frac{1}{4}}\right)\left(1+\sqrt{k}\right)^{\frac{1}{4}}k^{\frac{1}{16}}$$

Equation Eighteen

<u>Trigonometric Development of the Approximation</u>

It can also be demonstrated (as I have elsewhere) that a trigonometrical resolution of Equation Twelve yields the simplification:-

$$AGM_{(2)} = \sqrt{2^{-\frac{5}{2}}.\left[\left(\frac{v}{\sin(\theta)}\right)+2\left(\frac{v}{2}\right)\right].\left(\frac{v}{\sin(\theta)}\right)^{\frac{1}{4}}.\left(\sqrt{xy}\right)^{\frac{1}{4}}}$$

Equation Nineteen

where as before x = 1 and y = sin(α).
v is defined by:-

$$v = \sin(\alpha).(x+y)$$
$$= \sqrt{(x+y)^2 - (x-y)^2}$$
$$= \sqrt{(1+\sin(\alpha))^2 - (1-\sin(\alpha))^2}$$
$$= 2\sqrt{\sin(\alpha)}$$

Equation Twenty

whilst:-

$$\theta = \cos^{-1}\left(\frac{x-y}{x+y}\right)$$

Equation Twenty-One

so that:-

$$\sin(\theta) = \sin\left[\cos^{-1}\left(\frac{x-y}{x+y}\right)\right] = \sin\left[\cos^{-1}\left(\frac{1-\sin(\alpha)}{1+\sin(\alpha)}\right)\right] = \frac{2\sqrt{\sin(\alpha)}}{1+\sin(\alpha)}$$

Equation Twenty-Two

Substitution then permits us to re-write Equation Nineteen as:-

$$AGM_{(2)} = \sqrt{2^{-2.5}.\left[\left[\left(\frac{\dfrac{2\sqrt{\sin(\alpha)}}{2\sqrt{\sin(\alpha)}}}{1+\sin(\alpha)}\right) + 2\left(\frac{2\sqrt{\sin(\alpha)}}{2}\right)\right]\left(\frac{\dfrac{2\sqrt{\sin(\alpha)}}{2\sqrt{\sin(\alpha)}}}{1+\sin(\alpha)}\right)^{\frac{1}{4}}.\left(\sqrt{\sin(\alpha)}\right)^{\frac{1}{4}}\right]}$$

Equation Twenty-Three

Appropriate cancellations yield:-

$$AGM_{(2)} = \sqrt{2^{-2.5}.\left[(1+\sin(\alpha)) + 2\sqrt{\sin(\alpha)}\right]}.(1+\sin(\alpha))^{\frac{1}{4}}.\left(\sqrt{\sin(\alpha)}\right)^{\frac{1}{4}}$$

Equation Twenty-Four

which reduces to:-

$$AGM_{(2)} = 2^{-\frac{5}{4}}.\left(1+\sqrt{\sin(\alpha)}\right).(1+\sin(\alpha))^{\frac{1}{4}}.\left(\sqrt{\sin(\alpha)}\right)^{\frac{1}{4}}$$

Equation Twenty-Five

The substitution $\sqrt{k} = \sin(\alpha)$ then gives:-

$$AGM_{(2)} = 2^{-\frac{5}{4}}.\left(1+k^{\frac{1}{4}}\right).\left(1+\sqrt{k}\right)^{\frac{1}{4}}.k^{\frac{1}{16}}$$

Equation Twenty-Six

Equations Eighteen and Twenty-Six are of course identical.

Either of them gives a specific defect of –0.00000000486462 relative to the full AGM, that is to say ten-figure accuracy.

<u>The Use of Closed-Form AGM Approximations</u>

By combining Equations Seven and Eight the Complete Elliptic Integral of the First Kind, K(k), may further be defined as:-

$$K(k) = \frac{1}{2.AGM(1, \sin(\alpha))} = \frac{1}{2.AGM(1, \sqrt{k})}$$

Equation Twenty-Seven

This furnishes what is possibly the shortest and most efficient route to ultra-accuracy CEIFs, for which there are of course no exact values.

By substitution of Equation Twenty-Five in Equation Twenty-Seven followed by simplifications we are able to write:-

$$K(k) = \frac{\pi}{2.AGM} \approx \frac{\pi}{2\left[2^{-\frac{5}{4}}.\left(1+\sqrt{\sin(\alpha)}\right).\left(1+\sin(\alpha)\right)^{\frac{1}{4}}.\left(\sqrt{\sin(\alpha)}\right)^{\frac{1}{4}}\right]}$$

$$\approx \frac{\pi}{\left[2^{-\frac{1}{4}}.\left(1+\sqrt{\sin(\alpha)}\right).\left(1+\sin(\alpha)\right)^{\frac{1}{4}}.\left(\sqrt{\sin(\alpha)}\right)^{\frac{1}{4}}\right]}$$

Equation Twenty-Eight

Equation Twenty-Eight rearranges to form:-

$$K(k) \approx \frac{\sqrt[4]{2}.\pi}{\left(1+\sqrt{\sin(\alpha)}\right).\left(1+\sin(\alpha)\right)^{\frac{1}{4}}.\left(\sqrt{\sin(\alpha)}\right)^{\frac{1}{4}}} \approx \frac{\sqrt[4]{2}.\pi}{\left(1+k^{\frac{1}{4}}\right).\left(1+\sqrt{k}\right)^{\frac{1}{4}}.k^{\frac{1}{16}}}$$

Equation Twenty-Nine

For K(½) the specific defect relative to the full AGM value of this CEIF is 0.000000000048646, that is to say ten-figure accuracy, as per the AGM approximation.

<u>Some Computational Structures</u>

Using an EXCEL® spreadsheet I compared the Hastings Polynomial and three of my own identities against a five-pass AGM CEIF for 101 values k = {0,0.01,1}.

For all 101 values of the five-pass AGM the residual error c_5 was smaller than $\pm 10^{-16}$, except for k = 0 for which it was 0.015625.

My own three identities elaborated were:-

$$K_{warren1}(k) = \frac{2^{\frac{1}{4}}.\pi}{\left(1+k^{\frac{1}{4}}\right).\left(1+\sqrt{k}\right)^{\frac{1}{4}}.k^{\frac{1}{16}}} \qquad \textbf{Equation Thirty a}$$

$$K_{warren2}(k) = \frac{2^{\frac{1}{4}} . \pi}{\left(1 + \sqrt{k^{\frac{1}{2}}}\right) . \left(1 + k^{\frac{1}{2}}\right)^{\frac{1}{4}} . \left(k^{\frac{1}{2}}\right)^{\frac{1}{8}}}$$

Equation Thirty b

$$K_{warren3}(k) = \frac{\sqrt{\sqrt{2}} . \pi}{\left(1 + \sqrt{\sqrt{k}}\right) . \sqrt{\sqrt{1 + \sqrt{k}}} . \sqrt{\sqrt{\sqrt{\sqrt{k}}}}}$$

Equation Thirty c

For the AGM, Hastings and Warren1 relations a tabular summary is provided in Appendix One.

Figure One is a comparative plot of the Radix-16 logarithms of the absolute specific defects (relative to AGM) of the Hastings and Warren1 approximators.

It is manifest that though the Hastings Polynomial is cyclically unstable it is a superior estimator of CEIF from k = 0 to k = 0.5, when the Warren approximator becomes preferable.

From the point of view of verity the Warren2 and Warren3 structures are not significantly different to Warren1, though all three become numerically unstable at ~k>0.86 although not necessarily with defects coincident at the same arguments.

Such relations are illustrated by Figure Two.

It is clear from Equation Thirty c that the accuracy of the Warren estimator is driven by the precision and efficiency of whatever engine can be applied to the determination of repeated square roots.

<u>Root Mean Square Errors</u>

If u_i is some Estimate of a Particular CEIF and v_i is the relevant fiducial (i.e. full AGM) value at that point, then the Root Mean Square Error, RMS_{uv}, is given by:-

$$RMS_{uv} = \sqrt{\frac{\sum_{j=m}^{n} \left(\frac{u_i - v_i}{v_i}\right)^2}{(n - m + 1)}}$$

Equation Thirty-One

The smaller the absolute value of RMS_{uv} the better.

Often it is simpler to compare the absolute logarithm of RMS when dealing with very tiny discrepancies. For example:-

$$RMS_{\log 10} = \left| \log_{10} \sqrt{\frac{\sum_{j=m}^{n} \left(\dfrac{u_i - v_i}{v_i} \right)^2}{(n - m + 1)}} \right|$$

Equation Thirty-Two

The bigger $RMS_{\log 10}$ the better.

Table Two shows the overall superiority of Hastings to Warren1 with the latter having an RMS error more than three orders of magnitude greater than the former.

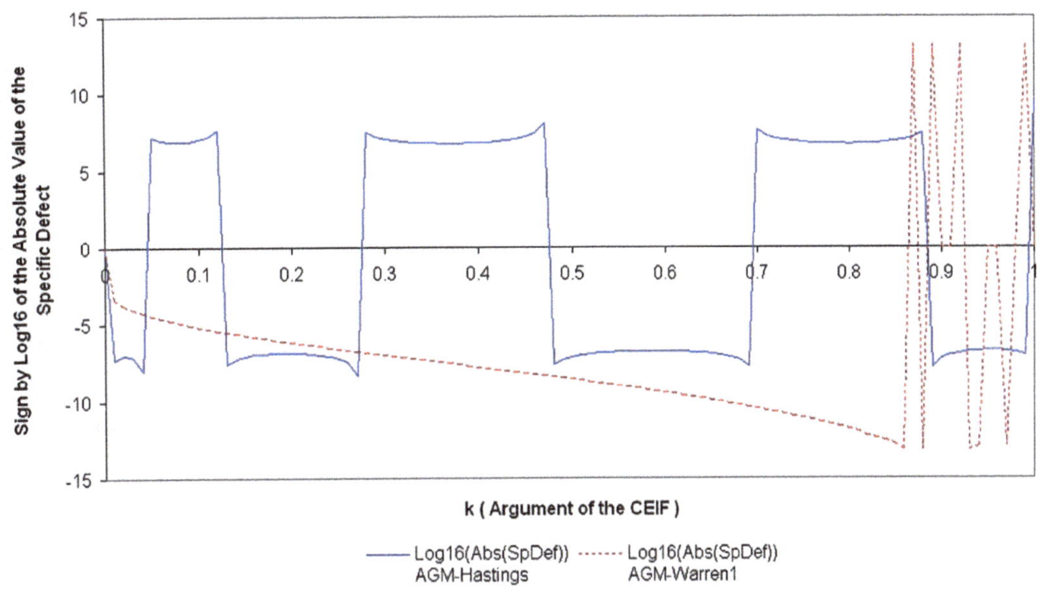

Figure One
Hastings and Warren1 Imprecisions
Relative to Fiducial Five-Pass AGMs

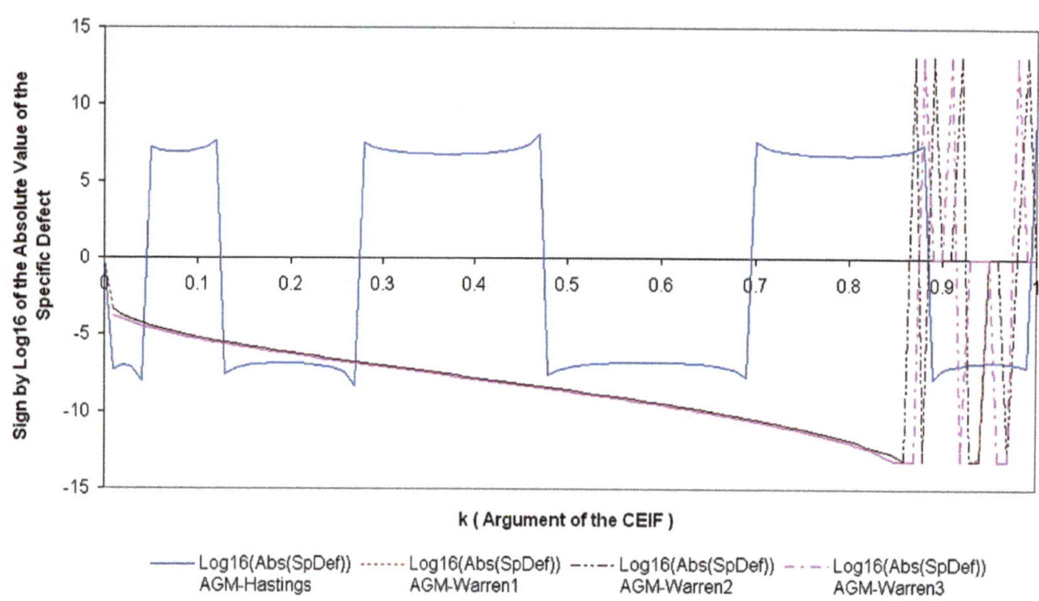

Figure Two
The Imprecisions of
Hastings and All Three Warren Estimators
Relative to Fiducial Five-Pass AGMs

Lower Bound:	0
Upper Bound:	1
Number of Intervals:	100
Interval:	0.01
RMS AGM-Hastings:	5.01914E-09
RMS AGM-Warren$_1$:	9.53785E-06

Table Two
Hastings and Warren1 RMS Errors

There is, however, a distinct reversal at k = 0.5 suggesting that the Warren estimator may be superior for K'(k), and this high-k benefit is shown in relief in Table Three.

		RMS	
		Hastings	**Warren1**
Series	**k**		
0-50	**0-0.5**	8.3865	4.8743
50-100	**0.5-1**	8.2428	10.9537

Table Three
Hastings and Warren1
Absolute Logarithm of RMS Errors

<u>References</u>

1   "Handbook of Mathematical Functions"
Milton Abramovitz and Irene A Stegun:
Dover Publications Incorporated   1964
SBN 486-61272-4
1046 pp (Softcover)

    Exact Formulas: 17.3.1 on Page 590
    Series Solution: 17.3.11 on Page 591
    Hastings Approximation: 17.3.34 on Page 591
    AGM Relations: Section 17.6. (Page 598)

2   "Approximations for Digital Computers"
Cecil Hastings Jnr
with Jeanne Hayward and James Wong
Princeton University Press of Princeton, New Jersey
ISBN 978-0691079141
December 1955
212 pages (Hardcover)
A research study by the RAND Corporation

3   "The Invention of a General Method for Determining the Sum
  of Every 2d, 3d, 4th, or 5th, &c. Term of a Series, Taken in Order:
  The Sum of the Whole Series Being Known"
Thomas Simpson FRS
The Philosophical Transactions of the Royal Society (1683-1775)
V50 (1757-1758) pp 757-769

4   "Gauss, Landen, Ramanujan, the Arithmetic-Geometric Mean,
  Ellipses, π, and the Ladies Diary"
Gert Almkvist and Bruce Berndt
The American Mathematical Monthly
V95 No 7 August-September 1988
pp 585-608
http://www.jstor.org/stable/2323302

    Gauss-Lagrange Formula on Page 588

5   "Elliptic integrals, the arithmetic geometric mean
 and the Brent-Salamin algorithm for π"
GJO Jameson
http://www.maths.lancs.ac.uk/~jameson/ellagm.pdf
    AGM relations: Section 3.6 Page 9

APPENDIX ONE
A Tabular Summary of Certain Estimates of Complete Elliptic Integrals of the First Kind

Serial	k	$k^{0.5}$	α	AGM CEIF	Hastings CEIF	Warren CEIF$_1$	Warren CEIF$_2$	Fractional Specific Defect AGM-Hastings	Specific Defect Squared AGM-Hastings	Fractional Specific Defect AGM-Warren	Specific Defect Squared AGM-Warren1
0	0	0	0								
1	0.01	0.1	0.100167421	3.695637363	3.695637369	3.695967573	3.695967573	1.75035E-09	3.06374E-18	8.93514E-05	7.98366E-09
2	0.02	0.141421356	0.141897055	3.354141446	3.354141459	3.354239199	3.354239199	3.99416E-09	1.59533E-17	2.91441E-05	8.4938E-10
3	0.03	0.173205081	0.174083011	3.155874948	3.155874956	3.155917305	3.155917305	2.6985E-09	7.28188E-18	1.34218E-05	1.80143E-10
4	0.04	0.2	0.201357921	3.016112492	3.016112493	3.016134309	3.016134309	2.31076E-10	5.33963E-20	7.23345E-06	5.23228E-11
5	0.05	0.223606798	0.225513406	2.908337248	2.908337242	2.908349684	2.908349684	-2.14416E-09	4.5974E-18	4.27577E-06	1.82822E-11
6	0.06	0.244948974	0.247467063	2.820752497	2.820752486	2.820760079	2.820760079	-3.87396E-09	1.50076E-17	2.68793E-06	7.22498E-12
7	0.07	0.264575131	0.267763327	2.747073004	2.747072991	2.747077856	2.747077856	-4.79216E-09	2.29648E-17	1.76618E-06	3.11938E-12
8	0.08	0.282842712	0.286756552	2.683551406	2.683551393	2.683554627	2.683554627	-4.93277E-09	2.43322E-17	1.19997E-06	1.43993E-12
9	0.09	0.3	0.304692654	2.627773332	2.62777332	2.627775531	2.627775531	-4.42495E-09	1.95802E-17	8.36934E-07	7.00458E-13
10	0.1	0.316227766	0.321750554	2.578092113	2.578092104	2.57809365	2.57809365	-3.43421E-09	1.17938E-17	5.96181E-07	3.55432E-13
11	0.11	0.331662479	0.338065255	2.533334546	2.533334541	2.533335641	2.533335641	-2.1298E-09	4.53604E-18	4.3212E-07	1.86728E-13
12	0.12	0.346410162	0.353741606	2.492635323	2.492635322	2.492636115	2.492636115	-6.67288E-10	4.45273E-19	3.17786E-07	1.00988E-13
13	0.13	0.360555128	0.368862984	2.455338028	2.45533803	2.455338609	2.455338609	8.19888E-10	6.72216E-19	2.36596E-07	5.59775E-14
14	0.14	0.374165739	0.383497004	2.42093296	2.420932966	2.420933391	2.420933391	2.22369E-09	4.94479E-18	1.78013E-07	3.16887E-14
15	0.15	0.387298335	0.397699415	2.389016486	2.389016495	2.389016809	2.389016809	3.46165E-09	1.19831E-17	1.3516E-07	1.82683E-14
16	0.16	0.4	0.411516846	2.359263555	2.359263565	2.359263799	2.359263799	4.47537E-09	2.00289E-17	1.03439E-07	1.06997E-14
17	0.17	0.412310563	0.424988783	2.331408568	2.33140858	2.331408754	2.331408754	5.22798E-09	2.73318E-17	7.97132E-08	6.3542E-15
18	0.18	0.424264069	0.438149031	2.305231737	2.30523175	2.305231879	2.305231879	5.70138E-09	3.25057E-17	6.18046E-08	3.81981E-15
19	0.19	0.435889894	0.451026812	2.280549138	2.280549152	2.280549248	2.280549248	5.89316E-09	3.47293E-17	4.81776E-08	2.32108E-15
20	0.2	0.447213595	0.463647609	2.257205327	2.25720534	2.257205412	2.257205412	5.81373E-09	3.37995E-17	3.77341E-08	1.42386E-15
21	0.21	0.458257569	0.476033818	2.235067755	2.235067768	2.235067822	2.235067822	5.48362E-09	3.00701E-17	2.96791E-08	8.8085E-16
22	0.22	0.469041576	0.488205263	2.214022498	2.214022509	2.21402255	2.21402255	4.93095E-09	2.43142E-17	2.34309E-08	5.49009E-16
23	0.23	0.479583152	0.500179609	2.193970925	2.193970935	2.193970966	2.193970966	4.1892E-09	1.75494E-17	1.85595E-08	3.44454E-16
24	0.24	0.489897949	0.511972688	2.17482709	2.174827097	2.174827122	2.174827122	3.29532E-09	1.08591E-17	1.47439E-08	2.17384E-16
25	0.25	0.5	0.523598776	2.156515647	2.156515652	2.156515673	2.156515673	2.28798E-09	5.23486E-18	1.17432E-08	1.37902E-16
26	0.26	0.509901951	0.535070807	2.138970184	2.138970186	2.138970204	2.138970204	1.20618E-09	1.45487E-18	9.37446E-09	8.78805E-17
27	0.27	0.519615242	0.546400564	2.122131863	2.122131863	2.122131879	2.122131879	8.80284E-11	7.749E-21	7.49847E-09	5.62271E-17
28	0.28	0.529150262	0.557598827	2.10594832	2.105948318	2.105948333	2.105948333	-1.03022E-09	1.06135E-18	6.00831E-09	3.60998E-17
29	0.29	0.538516481	0.568675503	2.090372747	2.090372742	2.090372757	2.090372757	-2.11494E-09	4.47296E-18	4.82147E-09	2.32466E-17
30	0.3	0.547722558	0.57963974	2.075363135	2.075363129	2.075363143	2.075363143	-3.13576E-09	9.83299E-18	3.87398E-09	1.50077E-17
31	0.31	0.556776436	0.590500015	2.060881647	2.060881638	2.060881653	2.060881653	-4.066E-09	1.65324E-17	3.11598E-09	9.70931E-18
32	0.32	0.565685425	0.601264217	2.046894077	2.046894067	2.046894082	2.046894082	-4.88297E-09	2.38434E-17	2.50845E-09	6.29232E-18
33	0.33	0.574456265	0.611939715	2.033369409	2.033369398	2.033369413	2.033369413	-5.56814E-09	3.10042E-17	2.02073E-09	4.08336E-18
34	0.34	0.583095189	0.62253342	2.020279429	2.020279416	2.020279432	2.020279432	-6.10727E-09	3.72988E-17	1.62865E-09	2.6525E-18
35	0.35	0.591607978	0.633051836	2.007598398	2.007598385	2.007598401	2.007598401	-6.49033E-09	4.21243E-17	1.31307E-09	1.72415E-18
36	0.36	0.6	0.643501109	1.995302778	1.995302764	1.99530278	1.99530278	-6.71142E-09	4.50432E-17	1.05881E-09	1.12108E-18
37	0.37	0.608276253	0.653887062	1.98337098	1.983370966	1.983370981	1.983370981	-6.76864E-09	4.58145E-17	8.53789E-10	7.28955E-19
38	0.38	0.6164414	0.664215238	1.971783162	1.971783149	1.971783163	1.971783163	-6.66382E-09	4.44065E-17	6.88362E-10	4.73842E-19
39	0.39	0.6244998	0.674490928	1.960521044	1.960521032	1.960521045	1.960521045	-6.40229E-09	4.09893E-17	5.54821E-10	3.07826E-19
40	0.4	0.632455532	0.684719203	1.94956775	1.949567738	1.949567751	1.949567751	-5.99256E-09	3.59108E-17	4.46986E-10	1.99796E-19
41	0.41	0.640312424	0.694904938	1.938907665	1.938907655	1.938907666	1.938907666	-5.44598E-09	2.96586E-17	3.59895E-10	1.29524E-19
42	0.42	0.64807407	0.705052837	1.928526318	1.928526309	1.928526319	1.928526319	-4.77638E-09	2.28138E-17	2.89559E-10	8.38442E-20
43	0.43	0.655743852	0.715167456	1.918410269	1.918410261	1.91841027	1.91841027	-3.99974E-09	1.59979E-17	2.32763E-10	5.41786E-20
44	0.44	0.663324958	0.725253222	1.908547016	1.90854701	1.908547017	1.908547017	-3.13376E-09	9.82044E-18	1.86915E-10	3.49372E-20
45	0.45	0.670820393	0.735314453	1.89892491	1.898924906	1.898924911	1.898924911	-2.19748E-09	4.82892E-18	1.49922E-10	2.24765E-20
46	0.46	0.678232998	0.745355373	1.889533079	1.889533077	1.889533079	1.889533079	-1.21093E-09	1.46635E-18	1.20091E-10	1.44219E-20
47	0.47	0.68556546	0.755380134	1.88036136	1.880361359	1.88036136	1.88036136	-1.94719E-10	3.79154E-20	9.60547E-11	9.2265E-21
48	0.48	0.692820323	0.765392826	1.87140024	1.871400241	1.87140024	1.87140024	8.30324E-10	6.89438E-19	7.67049E-11	5.88365E-21
49	0.49	0.7	0.775397497	1.862640802	1.862640806	1.862640802	1.862640802	1.84351E-09	3.39853E-18	6.11444E-11	3.73863E-21
50	0.5	0.707106781	0.785398163	1.854074677	1.854074683	1.854074677	1.854074677	2.82463E-09	7.97854E-18	4.86462E-11	2.36645E-21

Table A1
CEIF Estimates
Between k = 0 and k = 0.5

<table>
<tr><th rowspan="2">Serial</th><th rowspan="2">k</th><th rowspan="2">$k^{0.5}$</th><th rowspan="2">α</th><th>AGM</th><th>Hastings</th><th>Warren</th><th>Warren</th><th>Fractional
Specific
Defect
AGM-Hastings</th><th>Specific
Defect
Squared
AGM-Hastings</th><th>Fractional
Specific
Defect
AGM-Warren</th><th>Specific
Defect
Squared
AGM-Warren1</th></tr>
<tr><th>CEIF</th><th>CEIF</th><th>$CEIF_1$</th><th>$CEIF_2$</th></tr>
<tr><td>50</td><td>0.5</td><td>0.707106781</td><td>0.785398163</td><td>1.854074677</td><td>1.854074683</td><td>1.854074677</td><td>1.854074677</td><td>2.82463E-09</td><td>7.97854E-18</td><td>4.86462E-11</td><td>2.36645E-21</td></tr>
<tr><td>51</td><td>0.51</td><td>0.714142843</td><td>0.79539883</td><td>1.845693998</td><td>1.845694005</td><td>1.845693998</td><td>1.845693998</td><td>3.75429E-09</td><td>1.40947E-17</td><td>3.86214E-11</td><td>1.49161E-21</td></tr>
<tr><td>52</td><td>0.52</td><td>0.721110255</td><td>0.805403501</td><td>1.837491363</td><td>1.837491372</td><td>1.837491363</td><td>1.837491363</td><td>4.6142E-09</td><td>2.12909E-17</td><td>3.05931E-11</td><td>9.35939E-22</td></tr>
<tr><td>53</td><td>0.53</td><td>0.728010989</td><td>0.815416193</td><td>1.829459799</td><td>1.829459808</td><td>1.829459799</td><td>1.829459799</td><td>5.38747E-09</td><td>2.90249E-17</td><td>2.41744E-11</td><td>5.84404E-22</td></tr>
<tr><td>54</td><td>0.54</td><td>0.734846923</td><td>0.825440953</td><td>1.821592727</td><td>1.821592738</td><td>1.821592727</td><td>1.821592727</td><td>6.05886E-09</td><td>3.67098E-17</td><td>1.90526E-11</td><td>3.63E-22</td></tr>
<tr><td>55</td><td>0.55</td><td>0.741619849</td><td>0.835481874</td><td>1.813883937</td><td>1.813883949</td><td>1.813883937</td><td>1.813883937</td><td>6.61502E-09</td><td>4.37585E-17</td><td>1.49732E-11</td><td>2.24196E-22</td></tr>
<tr><td>56</td><td>0.56</td><td>0.748331477</td><td>0.845543105</td><td>1.806327559</td><td>1.806327572</td><td>1.806327559</td><td>1.806327559</td><td>7.04467E-09</td><td>4.96274E-17</td><td>1.17318E-11</td><td>1.37635E-22</td></tr>
<tr><td>57</td><td>0.57</td><td>0.754983444</td><td>0.855628871</td><td>1.798918039</td><td>1.798918052</td><td>1.798918039</td><td>1.798918039</td><td>7.33882E-09</td><td>5.38582E-17</td><td>9.16287E-12</td><td>8.39583E-23</td></tr>
<tr><td>58</td><td>0.58</td><td>0.761577311</td><td>0.86574349</td><td>1.791650117</td><td>1.79165013</td><td>1.791650117</td><td>1.791650117</td><td>7.49086E-09</td><td>5.6113E-17</td><td>7.13135E-12</td><td>5.08562E-23</td></tr>
<tr><td>59</td><td>0.59</td><td>0.768114575</td><td>0.875891389</td><td>1.784518805</td><td>1.784518818</td><td>1.784518805</td><td>1.784518805</td><td>7.49672E-09</td><td>5.62008E-17</td><td>5.53021E-12</td><td>3.05833E-23</td></tr>
<tr><td>60</td><td>0.6</td><td>0.774596669</td><td>0.886077124</td><td>1.777519371</td><td>1.777519385</td><td>1.777519371</td><td>1.777519371</td><td>7.35494E-09</td><td>5.40951E-17</td><td>4.27145E-12</td><td>1.82453E-23</td></tr>
<tr><td>61</td><td>0.61</td><td>0.781024968</td><td>0.896305399</td><td>1.770647323</td><td>1.770647336</td><td>1.770647323</td><td>1.770647323</td><td>7.06669E-09</td><td>4.99381E-17</td><td>3.28531E-12</td><td>1.07933E-23</td></tr>
<tr><td>62</td><td>0.62</td><td>0.787400787</td><td>0.906581089</td><td>1.763898389</td><td>1.763898401</td><td>1.763898389</td><td>1.763898389</td><td>6.63584E-09</td><td>4.40343E-17</td><td>2.51564E-12</td><td>6.32846E-24</td></tr>
<tr><td>63</td><td>0.63</td><td>0.793725393</td><td>0.916909265</td><td>1.757268505</td><td>1.757268516</td><td>1.757268505</td><td>1.757268505</td><td>6.0689E-09</td><td>3.68315E-17</td><td>1.9171E-12</td><td>3.67528E-24</td></tr>
<tr><td>64</td><td>0.64</td><td>0.8</td><td>0.927295218</td><td>1.750753803</td><td>1.750753812</td><td>1.750753803</td><td>1.750753803</td><td>5.37499E-09</td><td>2.88905E-17</td><td>1.45358E-12</td><td>2.11288E-24</td></tr>
<tr><td>65</td><td>0.65</td><td>0.806225775</td><td>0.93774449</td><td>1.744350597</td><td>1.744350605</td><td>1.744350597</td><td>1.744350597</td><td>4.56576E-09</td><td>2.08462E-17</td><td>1.096E-12</td><td>1.20121E-24</td></tr>
<tr><td>66</td><td>0.66</td><td>0.81240384</td><td>0.948262907</td><td>1.738055373</td><td>1.73805538</td><td>1.738055373</td><td>1.738055373</td><td>3.65526E-09</td><td>1.33609E-17</td><td>8.21718E-13</td><td>6.7522E-25</td></tr>
<tr><td>67</td><td>0.67</td><td>0.818535277</td><td>0.958856612</td><td>1.731864778</td><td>1.731864783</td><td>1.731864778</td><td>1.731864778</td><td>2.65977E-09</td><td>7.07438E-18</td><td>6.12209E-13</td><td>3.748E-25</td></tr>
<tr><td>68</td><td>0.68</td><td>0.824621125</td><td>0.96953211</td><td>1.72577561</td><td>1.725775612</td><td>1.72577561</td><td>1.72577561</td><td>1.59766E-09</td><td>2.55253E-18</td><td>4.53025E-13</td><td>2.05231E-25</td></tr>
<tr><td>69</td><td>0.69</td><td>0.830662386</td><td>0.980296312</td><td>1.719784808</td><td>1.719784809</td><td>1.719784808</td><td>1.719784808</td><td>4.89136E-10</td><td>2.39254E-19</td><td>3.32721E-13</td><td>1.10703E-25</td></tr>
<tr><td>70</td><td>0.7</td><td>0.836660027</td><td>0.991156586</td><td>1.713889448</td><td>1.713889447</td><td>1.713889448</td><td>1.713889448</td><td>-6.44033E-10</td><td>4.14778E-19</td><td>2.42529E-13</td><td>5.88202E-26</td></tr>
<tr><td>71</td><td>0.71</td><td>0.842614977</td><td>1.002120823</td><td>1.708086731</td><td>1.708086728</td><td>1.708086731</td><td>1.708086731</td><td>-1.77877E-09</td><td>3.16402E-18</td><td>1.75365E-13</td><td>3.07528E-26</td></tr>
<tr><td>72</td><td>0.72</td><td>0.848528137</td><td>1.0131975</td><td>1.702373977</td><td>1.702373972</td><td>1.702373977</td><td>1.702373977</td><td>-2.89104E-09</td><td>8.35809E-18</td><td>1.25867E-13</td><td>1.58426E-26</td></tr>
<tr><td>73</td><td>0.73</td><td>0.854400375</td><td>1.024395763</td><td>1.69674862</td><td>1.696748613</td><td>1.69674862</td><td>1.69674862</td><td>-3.95619E-09</td><td>1.56515E-17</td><td>8.93806E-14</td><td>7.9889E-27</td></tr>
<tr><td>74</td><td>0.74</td><td>0.860232527</td><td>1.03572552</td><td>1.691208199</td><td>1.691208191</td><td>1.691208199</td><td>1.691208199</td><td>-4.9494E-09</td><td>2.44966E-17</td><td>6.27583E-14</td><td>3.9386E-27</td></tr>
<tr><td>75</td><td>0.75</td><td>0.866025404</td><td>1.047197551</td><td>1.685750355</td><td>1.685750345</td><td>1.685750355</td><td>1.685750355</td><td>-5.84602E-09</td><td>3.4176E-17</td><td>4.37306E-14</td><td>1.91236E-27</td></tr>
<tr><td>76</td><td>0.76</td><td>0.871779789</td><td>1.058823639</td><td>1.680372823</td><td>1.680372812</td><td>1.680372823</td><td>1.680372823</td><td>-6.62213E-09</td><td>4.38526E-17</td><td>2.99958E-14</td><td>8.99748E-28</td></tr>
<tr><td>77</td><td>0.77</td><td>0.877496439</td><td>1.070616718</td><td>1.675073429</td><td>1.675073417</td><td>1.675073429</td><td>1.675073429</td><td>-7.25496E-09</td><td>5.26344E-17</td><td>2.02814E-14</td><td>4.11335E-28</td></tr>
<tr><td>78</td><td>0.78</td><td>0.883176087</td><td>1.082591063</td><td>1.669850086</td><td>1.669850073</td><td>1.669850086</td><td>1.669850086</td><td>-7.72347E-09</td><td>5.96519E-17</td><td>1.35632E-14</td><td>1.83961E-28</td></tr>
<tr><td>79</td><td>0.79</td><td>0.888819442</td><td>1.094762509</td><td>1.664700786</td><td>1.664700773</td><td>1.664700786</td><td>1.664700786</td><td>-8.00889E-09</td><td>6.41423E-17</td><td>8.93673E-15</td><td>7.98652E-29</td></tr>
<tr><td>80</td><td>0.8</td><td>0.894427191</td><td>1.107148718</td><td>1.659623599</td><td>1.659623585</td><td>1.659623599</td><td>1.659623599</td><td>-8.09533E-09</td><td>6.55344E-17</td><td>5.75306E-15</td><td>3.30977E-29</td></tr>
<tr><td>81</td><td>0.81</td><td>0.9</td><td>1.119769515</td><td>1.654616668</td><td>1.654616654</td><td>1.654616668</td><td>1.654616668</td><td>-7.9704E-09</td><td>6.35274E-17</td><td>3.75752E-15</td><td>1.41189E-29</td></tr>
<tr><td>82</td><td>0.82</td><td>0.905538514</td><td>1.132647296</td><td>1.649678205</td><td>1.649678193</td><td>1.649678205</td><td>1.649678205</td><td>-7.62589E-09</td><td>5.81541E-17</td><td>1.88438E-15</td><td>3.5509E-30</td></tr>
<tr><td>83</td><td>0.83</td><td>0.911043358</td><td>1.145807544</td><td>1.644806491</td><td>1.644806479</td><td>1.644806491</td><td>1.644806491</td><td>-7.05841E-09</td><td>4.98211E-17</td><td>1.21498E-15</td><td>1.47617E-30</td></tr>
<tr><td>84</td><td>0.84</td><td>0.916515139</td><td>1.159279481</td><td>1.639999866</td><td>1.639999856</td><td>1.639999866</td><td>1.639999866</td><td>-6.27019E-09</td><td>3.93153E-17</td><td>8.12358E-16</td><td>6.59926E-31</td></tr>
<tr><td>85</td><td>0.85</td><td>0.921954446</td><td>1.173096912</td><td>1.635256732</td><td>1.635256724</td><td>1.635256732</td><td>1.635256732</td><td>-5.26979E-09</td><td>2.77707E-17</td><td>4.07357E-16</td><td>1.6594E-31</td></tr>
<tr><td>86</td><td>0.86</td><td>0.92736185</td><td>1.187299323</td><td>1.630575549</td><td>1.630575542</td><td>1.630575549</td><td>1.630575549</td><td>-4.07291E-09</td><td>1.65886E-17</td><td>1.36176E-16</td><td>1.85438E-32</td></tr>
<tr><td>87</td><td>0.87</td><td>0.932737905</td><td>1.201933343</td><td>1.625954829</td><td>1.625954825</td><td>1.625954829</td><td>1.625954829</td><td>-2.70317E-09</td><td>7.30714E-18</td><td>-1.36563E-16</td><td>1.86493E-32</td></tr>
<tr><td>88</td><td>0.88</td><td>0.938083152</td><td>1.217054721</td><td>1.621393138</td><td>1.621393136</td><td>1.621393138</td><td>1.621393138</td><td>-1.19301E-09</td><td>1.42328E-18</td><td>1.36947E-16</td><td>1.87544E-32</td></tr>
<tr><td>89</td><td>0.89</td><td>0.943398113</td><td>1.232731072</td><td>1.616889091</td><td>1.616889091</td><td>1.616889091</td><td>1.616889091</td><td>4.15452E-10</td><td>1.726E-19</td><td>-1.37328E-16</td><td>1.88591E-32</td></tr>
<tr><td>90</td><td>0.9</td><td>0.948683298</td><td>1.249045772</td><td>1.612441349</td><td>1.612441352</td><td>1.612441349</td><td>1.612441349</td><td>2.06954E-09</td><td>4.28298E-18</td><td>0</td><td>0</td></tr>
<tr><td>91</td><td>0.91</td><td>0.953939201</td><td>1.266103673</td><td>1.60804862</td><td>1.608048626</td><td>1.60804862</td><td>1.60804862</td><td>3.70502E-09</td><td>1.37272E-17</td><td>0</td><td>0</td></tr>
<tr><td>92</td><td>0.92</td><td>0.959166305</td><td>1.284039775</td><td>1.603709655</td><td>1.603709663</td><td>1.603709655</td><td>1.603709655</td><td>5.24521E-09</td><td>2.75123E-17</td><td>-1.38457E-16</td><td>1.91703E-32</td></tr>
<tr><td>93</td><td>0.93</td><td>0.964365076</td><td>1.303033</td><td>1.599423245</td><td>1.599423255</td><td>1.599423245</td><td>1.599423245</td><td>6.59992E-09</td><td>4.35589E-17</td><td>1.38828E-16</td><td>1.92732E-32</td></tr>
<tr><td>94</td><td>0.94</td><td>0.969535971</td><td>1.323329264</td><td>1.595188221</td><td>1.595188234</td><td>1.595188221</td><td>1.595188221</td><td>7.66445E-09</td><td>5.87438E-17</td><td>2.78393E-16</td><td>7.75027E-32</td></tr>
<tr><td>95</td><td>0.95</td><td>0.974679434</td><td>1.345282921</td><td>1.591003454</td><td>1.591003467</td><td>1.591003454</td><td>1.591003454</td><td>8.31856E-09</td><td>6.91984E-17</td><td>0</td><td>0</td></tr>
<tr><td>96</td><td>0.96</td><td>0.979795897</td><td>1.369438406</td><td>1.586867847</td><td>1.586867861</td><td>1.586867847</td><td>1.586867847</td><td>8.42536E-09</td><td>7.09867E-17</td><td>0</td><td>0</td></tr>
<tr><td>97</td><td>0.97</td><td>0.98488578</td><td>1.396713316</td><td>1.582780342</td><td>1.582780355</td><td>1.582780342</td><td>1.582780342</td><td>7.83023E-09</td><td>6.13125E-17</td><td>2.80575E-16</td><td>7.87225E-32</td></tr>
<tr><td>98</td><td>0.98</td><td>0.989949494</td><td>1.428899272</td><td>1.578739912</td><td>1.578739922</td><td>1.578739912</td><td>1.578739912</td><td>6.35966E-09</td><td>4.04453E-17</td><td>0</td><td>0</td></tr>
<tr><td>00</td><td>0.99</td><td>0.994987437</td><td>1.470628906</td><td>1.574745562</td><td>1.574745568</td><td>1.574745562</td><td>1.574745562</td><td>3.82013E-09</td><td>1.45934E-17</td><td>-1.41003E-16</td><td>1.9882E-32</td></tr>
<tr><td>100</td><td>1</td><td>1</td><td>1</td><td>1.570796327</td><td>1.570796327</td><td>1.570796327</td><td>1.570796327</td><td>-3.11723E-12</td><td>9.7171E-24</td><td>0</td><td>0</td></tr>
</table>

Table A2
CEIF Estimates
Between k = 0.5 and k = 1

CHAPTER ONE

Stream Hydraulics

Aspects of the Design of Neutrally-Buoyant
Liquid Ballasted Tachometric Floats

by
James R Warren BSc MSc PhD PGCE

PART ONE
THE DETERMINATION OF STATIC MASSES

According to The Principle of Archimedes a buoyant object sits at the equilibrium of its downward weight with the upthrust due to the weight of fluid it displaces.

A neutrally-buoyant object can rest in the midst of a still pond of fluid because, in the presence of a common Local Acceleration Due to Gravity, its mass precisely matches that of the displaced fluid so that no unbalanced force tends either to raise or sink the object from where it happens to be.

That is to say:-

$$F_{up} = F_{down}$$

Equation 1

or:-

$$M_D g = M_{float} g$$
$$\therefore M_D = M_{float}$$

Equation 2

In practice of course exactitude is infeasible and the submerged object will tend slowly to rise or sink with some Stokesian[1] velocity conditioned by environmental factors.

Such emergent tendencies may be negligible in active hydraulic flows, and this disquisition is predicated upon the proposition that a neutrally-buoyant float has as higher probability of occupying any particular flow streamline as any other.

So though it is well-known that the rate of flow varies across the section of a bounded stream we may anticipate that a neutrally-buoyant float has a uniform probability of being at any point in the sectional area at a random time.

Therefore, should the length of a stream course be known, and the times of a float's upstream release and terminal capture, then a representative velocity local in space and time, and integrative of the typical section, may readily be computed.

This assumption can of course be tested in flumes against calibrated instruments, with and without obstructions and sediments. In real life actual floats will be subject to loss, breakage, capture and attrition.

Accordingly, we look for cheap and simple floats that can be developed and deployed by the amateur and the Third World practitioner who is making spot assessments of the discharge of wadeable or immature watercourses.

Ping-pong balls are manufactured of cellulose nitrate and delivered by the gross. The relevant international board of control for the sport of ping-pong specifies that each ball must be a hollow sphere of 40mm diameter; 27 grams mass; and have an elastic Coefficient of Restitution of 0.4.

Our model of float development envisions that a fine drill is used to puncture the spherical envelope (shell) and that, to simplify initial analysis, a dense liquid is instilled through to the chamber within. The amount of this ballast liquid is calculated with extreme precision to make the hydrostatic balance, and the hole then sealed with epoxy resin or some suitable cement.

The float is finished with a high visibility attrition-resistant marine paint and a printed or painted label of some unique identifier.

The insertion of ballast does of course partially displace the air or other fluid that occupied the interior of the ping-pong ball. If the requisite extreme precision of ballasting is to be attained than the decrement of displaced mass must be allowed against the increment of ballast mass to achieve a composite chamber target mass.

Therefore, the identification of this required mass necessitates the solution of a cubic equation[2]. We shall see that this system involves the geometry of the cap (spherical segment) and the sphere, and results in three real roots, two of which are, conveniently, absurd.

The second, valid, root can be used to compute a high-precision ballast target mass. It is notable that departures from the plane interface due to menisci are not germane to the computation as they do not affect the injectable ballast mass.

In the principal development the dense fluid ballast is assumed to be mercury though of course no modern conservancy would thank us for introducing this to their river, even in elemental form, and the costs and controls that the baleful metal would engage are unreasonable.

There is no reason in principle why any dense fluid that does not attack cellulose nitrate could not be used, such as a commercial mud for petroleum drilling, or a laboratory-made barium slurry.

The ballast can indeed be water, and a secondary development will illustrate the calculations involved in using that.

Figure One illustrates the vertical section through a mercury-ballasted float suspended in water, whilst Figure Two shows the same set-up with water ballast. In both pictures the thickness of the float shell is exaggerated. The menisci are also exaggerated and are for illustrative purposes only.

On the other hand, I have scaled the central depths of both ballast pools accurately to represent the necessary respective depths computed, as if meniscus formation did not condition that.

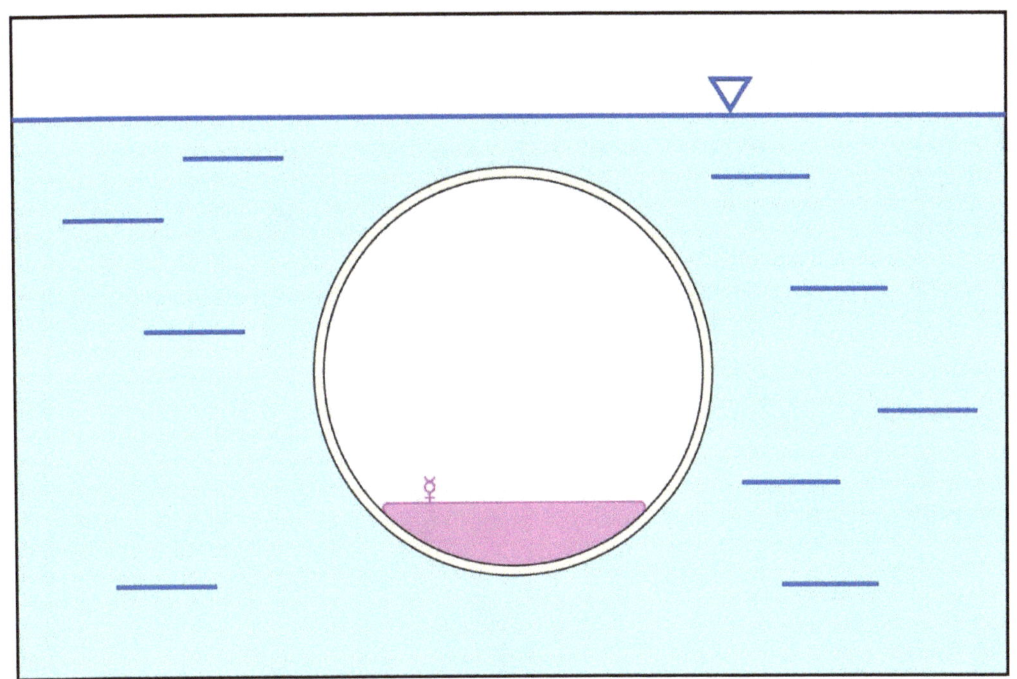

Figure One
A Neutrally-Buoyant Mercury Ballasted Float

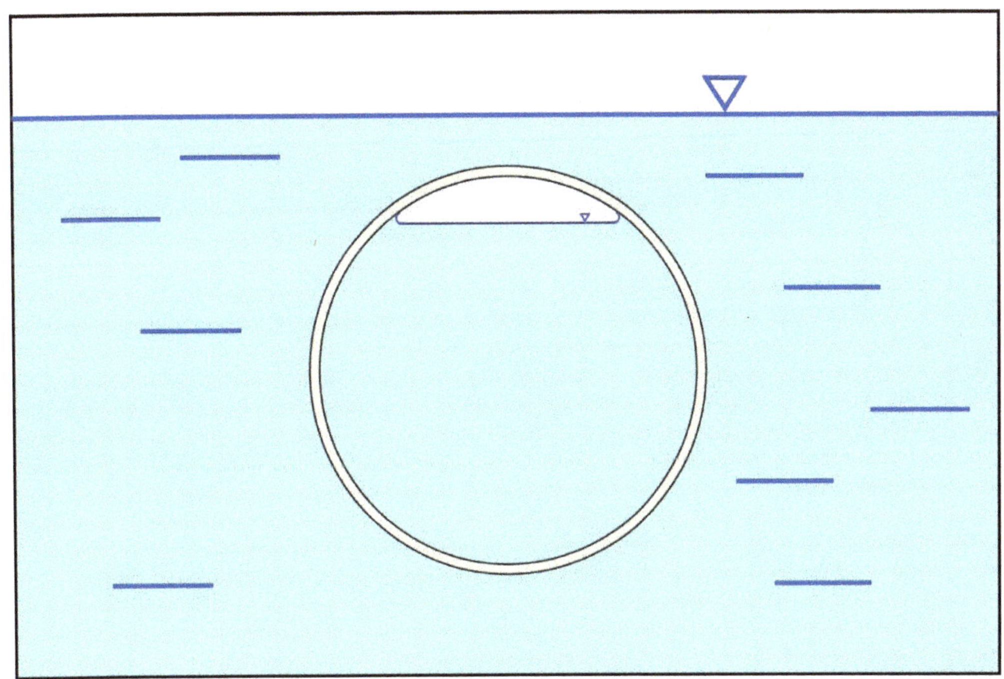

Figure Two
A Neutrally-Buoyant Water Ballasted Float

<u>Materials Notation</u>

Table One in Appendix One presents the Masses, Volumes and other principal quantities notations for selected components of the manufactured float.

<u>General Notation</u>

Table Two in Appendix One shows notation used in the mathematical development of the float manufacturing model, and other notation requisite.

<u>Notes upon Materials and Conditions</u>

Table Three in Appendix One lists some of the physical and commercial considerations that affected materials and components at Bloxwich, England in the Autumn of 2011.

<u>Densities</u>
The assumed densities of some selected materials are given in Table Four in Appendix One.

Equilibrium Diagram

The equilibrium diagram of Figure Three labels some of the basic notation applicable to this hydrostatic system.

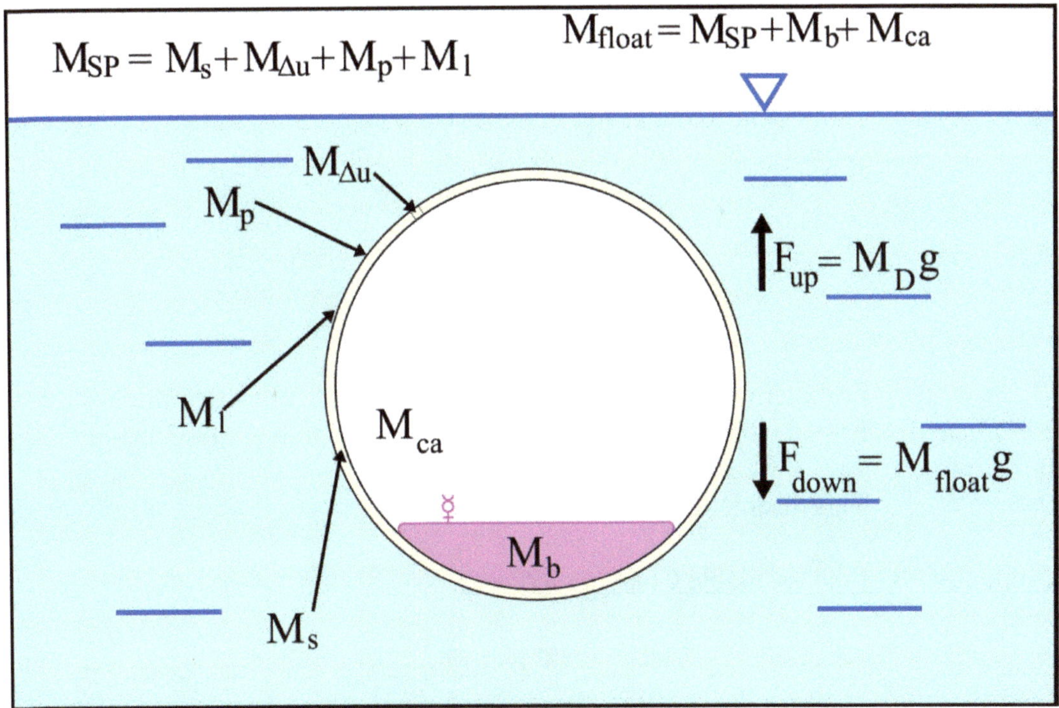

Figure Three
Equilibrium Diagram

<u>Computation of Shell Thickness</u>

The Shell Thickness, t_s, of the ping-pong ball float is given by:-

$$t_s = r_s - \sqrt[3]{r_s^{\,3} - \frac{3V_s}{4\pi}}$$

Equation 3

where:-

$$V_s = \frac{M_s}{\rho_s}$$

Equation 4

For international standard ping-pong balls of cellulose nitrate M_s = 0.0027 kg; ρ_s = 1375 kg/m³ and r_s = 0.02 meters. Accordingly, t_s computes to be 0.000398542079295 meters, call it 0.4 mm.

This disquisition is SI throughout.

<u>Computation of the Available Chamber Volume</u>

$$V_C = \frac{4}{3}\pi r_{ca}^{\,3}$$

Equation 5

<u>Computation of the Full-Cover Paint Volume, Mass and Area</u>

The Thickness of the dried marine paint coating is supposed to be t_p = 0.000125 meters or 125 microns conformable to the manufacturers' consensual assumptions.

I computed that the dried density of this material ρ_p = 1721.2 kg/m³.

Further:-

$$r_p = r_s + t_p$$

Equation 6

or in this case:- r_p = 0.020125 meters.

The dried Coat Volume V_p is given by:-

$$V_p = \frac{4}{3}\pi\left(r_s + t_p\right)^3 - \frac{4}{3}\pi r_s^{\,3}$$

Equation 7

$$V_p = \frac{4}{3}\pi t_p \left[3r_s \left(r_s + t_p \right) + t_p^{\,2} \right]$$
Equation 8

This computes to 0.000000632253703 m^3.
The Dried Paint Mass M$_p$ is therefore:-

$$M_p = \rho_p V_p$$
Equation 9

So M$_p$ = 0.001088235073201 kg.
The Area of this marine paint coat, called the Under Paint Area, UPA, is given by:-

$$UPA = 4\pi \left(r_s + t_p \right)^2$$
Equation 10

UPA = 0.005089576448356 m^2.

Computation of Label (Part-Cover) Paint Volume and Mass

In our example Label Text Area LTA is 0.0004 m^2 and is assumed to be paint of the same density and thickness as the underlying general coat.
Accordingly, the Fractional Label Area, f$_c$, is:-

$$f_c = \frac{LTA}{UPA}$$
Equation 11

whilst the Label Volume, V$_l$, is given by:-

$$V_l = t_p . LTA$$
Equation 12

and the Label Mass, M$_l$, is:-

$$M_l = \rho_p V_l = \rho_p . t_p . LTA$$
Equation 13

Computation of Plug Volume

The ballast is inserted through a hole drilled in the shell, which is then repaired with a cement.

In ideal terms the volume removed and replaced constitutes a cylinder modified by a shallow, additive upper cap and a deeper privative lower cap.

Of course, hand manufacture is very imperfect and in practice the lower cap is likely to be markedly concave and the upper absent.

Notwithstanding that, we shall continue our design analysis as if the restored bulk was perfect.

Figure Four illustrates the geometry in section.

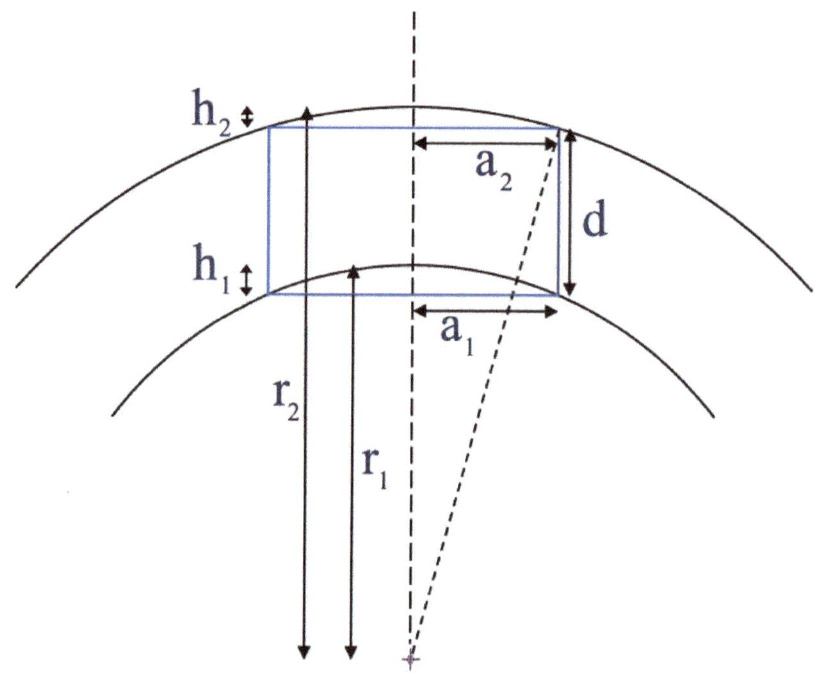

Figure Four
The Plug Section Annotated

The cap radii a$_1$ and a$_2$ identify with Drill Radius, r, which for exemplification we will declare to be 1 mm or r = 0.001 meters, perhaps an impracticably big hole.

Therefore:-

$$h_1 = r_{ca} - \sqrt{r_{ca}^2 - a^2} \qquad \textbf{Eqn. 14a}$$

$$h_2 = r_s - \sqrt{r_s^2 - a^2} \qquad \textbf{Eqn. 14b}$$

and the Plug Central Depth is:-

$$d_u = t_s + h_1 - h_2$$

Equation 15

From this it follows that:-

$$V_{cap1} = \frac{\pi h_1^{\,2}}{3}\left(3r_{ca} - h_1\right)$$

Equation 16

and:-

$$V_{cap2} = \frac{\pi h_2^{\,2}}{3}\left(3r_s - h_2\right)$$

Equation 17

whilst:-

$$V_{cyl} = \pi a^2 d_u$$

Equation 18

Therefore the Plug Volume, V_u, is:-

$$V_u = V_{cyl} + V_{cap2} - V_{cap1}$$

Equation 19

which expands to:-

$$V_u = \pi\left[a^2\left(t_s - h_2 + h_1\right) + h_2^{\,2}\left(r_s - \frac{h_2}{3}\right) - h_1^{\,2}\left(r_{ca} - \frac{h_1}{3}\right)\right]$$

Equation 20

Computation of the Plug Mass Differential

The corrective Plug Mass Differential, $M_{\Delta u}$, is given by:-

$$M_{\Delta u} = V_u\left(\rho_e - \rho_s\right)$$

Equation 21

ρ_e is the density of Dried Plug Cement that we will assume to be epoxy resin of density 975.58890717 kg/m^3.

Determination of Displaced Mass

The reactive Mass of Displaced Fluid, M_D, is controlled by the volume of the painted float incremented by a tiny amount V_l, the painted Label Volume.

The Displaced Mass is therefore:-

$$M_D = \rho_w \left(\frac{4}{3} \pi r_p^{\ 3} + V_l \right)$$

Equation 22

Determination of Carcase Composite Mass

The assembled Mass of the Shell, Plug and Body with the Label Paint, M_{SP}, is given by:-

$$M_{SP} = M_s + M_{\Delta u} + M_p + M_l$$

Equation 23

Determination of Chamber Content Target Mass

The Target Mass of the fluid and ballast within the float cavity; that Mass that gives hydrostatic equilibrium and Stokesian stasis (i.e. neutral buoyancy) to the entire float is computed by:-

$$M_T = M_D - M_{SP}$$

Equation 24

For our exemplary mercury:air balance system the quantity M_T is 0.030250395521892 if the Density of Mercury ρ_b is 13546 kg/m^3 and the Density of Contained Air is 1.2041 Kg/m^3.

The Determination of the Enclosed Masses

The meeting of M_T by instilling a ballast into the float cavity is controlled by the tandem Air and Ballast Masses M_{ca} and M_b.
In the case of an instilled liquid, modelled as a cap or spherical segment, M_b is only accessible via knowledge of the ballast fluid Central Depth, d.
The relationship between M_T and d is articulated by:-

$$M_T = M_{ca} + M_b = \rho_{ca} V_{ca} + \rho_b V_b$$

Equation 25

which may be expanded into the cubic equation:-

$$M_T = \rho_{ca}\left[\frac{4}{3}\pi r_{ca}^{\;3} - \frac{\pi d^2}{3}\left(3r_{ca} - d\right)\right] + \rho_b\left[\frac{\pi d^2}{3}\left(3r_{ca} - d\right)\right]$$

Equation 26

appropriate substitutions and re-arrangements configure this cubic equation as:-

$$M_T = \pi\left[r_{ca}d^2\left(\rho_b - \rho_{ca}\right) + \frac{d^3}{3}\left(\rho_{ca} - \rho_b\right) + \rho_{ca}\frac{4}{3}r_{ca}^{\;3}\right]$$

Equation 27

or in terms of Coefficients $k_0 \ldots k_3$:-

$$M_T = k_0 + k_1 d + k_2 d^2 + k_3 d^3$$

Equation 28

<u>The Computation of the Chamber Mass Differential Variables</u>

The Infill Differential Density, $\Delta\rho_\alpha$, is given by:-

$$\Delta\rho_\alpha = \rho_b - \rho_{ca}$$

Equation 29

and the Whole Chamber Mass Differential is:-

$$\Delta M_C = \Delta\rho_\alpha.V_C$$

Equation 30

where V_C is the Volume of the Empty Cavity, i.e.:-

$$V_C = \frac{4}{3}\pi r_{ca}^{\;3}$$

Equation 31

Therefore, the Mass of Air Contained in the Air-filled Chamber with No Ballast ("The Empty Chamber"), M_F, is yielded by:-

$$M_F = \rho_{ca}.V_C$$

Equation 32

In our mercury:air ballasting system $\Delta\rho_\alpha = 13544.7959$; $\Delta M_C = 0.427293413366734$ and $M_F = 0.000037985363739$.

PART TWO
THE SOLUTION FOR BALLAST DEPTH

Appropriate resolution of the terms of Equation Twenty-Seven shows that the four Coefficients of the cubic solution equation are:-

$$k_0 = M_F - M_T \qquad \textbf{Eqn.33a}$$

$$k_1 = 0 \qquad \textbf{Eqn.33b}$$

$$k_2 = \pi.\Delta\rho_\alpha.r_{ca} \qquad \textbf{Eqn.33c}$$

$$k_3 = \frac{-\pi.\Delta\rho_\alpha}{3} \qquad \textbf{Eqn.33d}$$

<u>The Areal Mass Density Cofactor</u>

The Areal Mass Density Cofactor, CF, is given by:-

$$CF = k_2{}^2 - 3k_1 k_3$$

Equation 34

For our system this reduces to:-

$$CF = k_2{}^2$$

Equation 35

or:-

$$CF = \left(\pi.\Delta\rho_\alpha.r_{ca}\right)^2$$

Equation 36

Which is, perforce, positive and non-zero.
Our CF is +695699.0736145069.
If CF is non-zero then three roots exist.

<u>The Discriminant Δ</u>

The Discriminant of a Cubic Equation, Δ, is given by:-

$$\Delta = 18k_0 k_1 k_2 k_3 - 4k_0 k_2{}^3 + k_1{}^2 k_2{}^2 - 4k_1{}^3 k_3 - 27k_0{}^2 k_3{}^2$$

Equation 37

For our system this reduces to:-

$$\Delta = -4k_0 k_2{}^3 - 27k_0{}^2 k_3{}^2$$

Equation 38

or:-

$$\Delta = 4(\pi.r_{ca}.\Delta\rho_\alpha)^3(M_T - M_F) - 27\left(\frac{-\Delta\rho_\alpha.\pi}{3}\right)^2(M_T - M_F)^2$$

Equation 39

Because M_T is much greater than M_F, Δ is almost inevitably positive.
The computed Δ for our set-up is +65,167,402.82307701.
Since Δ is positive there are three *real* roots to the cubic equation.
Each root constitutes a possible Central Ballast Depth, d, for us to choose from.
Which root we choose shall depend upon the physical realities of the situation. The justice of our choice can be checked by substitutive solutions.

<u>Arrangement of the Real Auxiliary, R</u>

The Cubic Equation Solution Auxiliary R is always real. It may be defined as:-

$$R = 2k_2{}^3 - 9k_1 k_2 k_3 + 27k_0 k_3{}^2$$

Equation 40

which on elimination of k_1 is:-

$$R = 2k_2{}^3 + 27k_0 k_3{}^2$$

Equation 41

In physical terms that is:-

$$R = 3\pi^2.\Delta\rho_\alpha\left[\frac{\Delta M_c}{2} + (M_F - M_T)\right]$$

Equation 42

Invariably an enormous positive number. In fact, our R for the mercury:air ping-pong float is +996,429,253.2942398.

<u>Arrangement of the Real Multiplier, q</u>

q is defined by:-

$$q = 3\pi^2.\Delta\rho_\alpha^{\,2}\sqrt{\Delta M_C + M_F - M_T}\sqrt{M_T - M_F}$$

Equation 43

Where the density of the ballast exceeds that of air q is invariably a large positive real.

Our q is +594,974,252.155915.

<u>Arrangement of the Imaginary Auxiliary, Q</u>

Because (M_F-M_T) is invariably negative and ΔM_C is much greater than (M_T-M_F), Q is always complex of form 0+yi.

In fact:-

$$Q = 0 + qi$$

Equation 44

<u>Arrangements of the Root Auxiliary, C</u>

The auxiliary C is invariably complex, being as it is the cube root of a complex number.

It is defined by:-

$$C = \sqrt[3]{\frac{Q+R}{2}}$$

Equation 45

In physical terms this may be written as:-

$$C = \sqrt[3]{\frac{3\left(\pi^2 \Delta\rho_\alpha^{\,2}\right)}{2}\left[\frac{\Delta M_C}{2} + \left(M_F - M_T\right) + \sqrt{\left(M_F - M_T\right)^2 + \Delta M_C}\right]}$$

Equation 46

In our mercury:air ping-pong ball float example system the numeric value of C is 820.6950517061698+148.85800522482239i.

<u>The Three Roots</u>

For this hydrostatic system the three (real) roots are offered as firstly:-

$$x_1 = \frac{k_2}{3k_3} - \frac{C}{3k_3} - \frac{CF}{3K_3.C}$$

Equation 47a

or:-

$$x_1 = r_{ca} + \frac{1}{\pi.\Delta\rho_\alpha}\left(C + \frac{CF}{C}\right)$$

Equation 47b

Secondly:-

$$x_2 = \frac{k_2}{3k_3} + \frac{C\left(1 + \sqrt{3}i\right)}{6k_3} + \frac{CF\left(1 - \sqrt{3}i\right)}{C.6k_3}$$

Equation 48a

or:-

$$x_2 = r_{ca} - \frac{1}{2\pi.\Delta\rho_\alpha}\left[C\left(1 + \sqrt{3}i\right) + \frac{CF.\left(1 - \sqrt{3}i\right)}{C}\right]$$

Equation 48b

and thirdly:-

$$x_3 = \frac{k_2}{3k_3} + \frac{C\left(1 - \sqrt{3}i\right)}{6k_3} + \frac{CF\left(1 + \sqrt{3}i\right)}{C.6k_3}$$

Equation 49a

or:-

$$x_3 = r_{ca} - \frac{1}{2\pi.\Delta\rho_\alpha}\left[C\left(1 - \sqrt{3}i\right) + \frac{CF.\left(1 + \sqrt{3}i\right)}{C}\right]$$

Equation 49b

<u>Root Values</u>

The values computed for these roots are physically tantamount to Ballast Depths within the hydrostatic float chamber.

The values for the mercury:air system and the given data are:-

$$x_1 = 0.058174995747169$$
$$x_2 = 0.006373821991182$$
$$x_3 = -0.005744443976236$$

The First Root, x_1, is absurd by reason of the 58cm length exceeding by some fourteen times the available diameter of the ping-pong ball float.

The Third Root, x_3, is absurd by reason of being negative, since negative depth (elevation) is not relevant to our context.

The only feasible root, The Second Root, x_2, offers a depth of 6.37 mm of mercury as appropriate to our objective of ballasting to neutral buoyancy.

Let us define the Central Ballast Depth, d, as:-

$$d = x_2$$

Equation 50

This quantity is the theoretical central height of the geometric cap of interest in computing the ballast mass. It may of course differ from the realised central depth due to the effects of surface tension (meniscus development).

<u>Validation of the Central Ballast Depth</u>

Allow $M_T{'}$ to be the value of Total Mass inferred by d.
Ballast Mass, M_b, is now computable using:-

$$M_b = \rho_b V_b = \rho_b \left[\frac{\pi d^2}{3} (3r_{ca} - d) \right]$$

Equation 51

whilst the Residual Mass of Contained Air, M_{ca}, is yielded by:-

$$M_{ca} = \rho_{ca} V_{ca} = \rho_{ca} \left[\frac{4}{3} \pi r_{ca}^3 - \frac{\pi d^2}{3} (3r_{ca} - d) \right]$$

Equation 52

or:-

$$M_{ca} = \frac{\pi . \rho_{ca}}{3} \left(4r_{ca}^3 - 3r_{ca} d^2 + d^2 \right)$$

Equation 53

$M_T{'}$ is then:-

$$M_T{}' = M_b + M_{ca}$$

Equation 54

The computed value of M_T' agrees exactly with the input M_T value at 0.030250395521892 (30.2504 grams).

PART THREE
SYSTEM SOLUTIONS

Solutions Transput Tableau (Mercury:Air Couple)

Table Five in Appendix Two presents the Solutions Tableau that I realised as an EXCEL® worksheet.

As per convention, input values are red and output blue.

Table Six offers some ancillary Results for the Paint, Plug and other desiderata.

The cubic equation solution intermediates are in Table Seven.

The background EXCEL® formulaic tableaus are presented in Appendix Three.

The Water:Air Ballast System

The outlines of the mercury:air system methodology are applicable to a neutrally-buoyant float ballasted with water.

In such a system it must be remembered that due to the low density of both fillers the water occupies the entire lower hemisphere and tops toward the crown of the float.

Therefore we may think of the residual air mathematically as cap-filling ballast and the water as the default chamber fluid. I.e. $\rho_b = 1.2041$ kg/m^3 and $\rho_{ca} = 998$kg/m^3.

The elaboration is essentially the same and the cubic roots for our specimen ping-pong ball float are:-

$$x_1 = -0.004325932536529$$
$$x_2 = 0.00467163260474$$
$$x_3 = 0.058458673693903$$

As with the mercury:air couple the first and third roots are absurd for complementary reasons, and the second root is of design interest.

The indicated air cap central depth is about 4.6716 mm, or in other words the ballast water body is around 35.5313 mm deep.

The Mass of Water Ballast, M_{ca}, is 0.030248905857093 kg and the Air Cap Mass, M_b, is 0.000001489664799 kg.

The relevant spreadsheet and solutions lists are shown in Appendix Four, whilst the background EXCEL® formulaic tableaus are presented in Appendix Five. The water:air formulae differ in elaborative detail from the mercury:air formulae.

Stokesian Sensitivity

The control of ballast mass is extremely critical if settling or emergence is to be suppressed. In practice, of course, true stasis is impossible if only because environmental temperatures and pressures are variable.

On the other hand wadeable and immature streams are likely to be so energetic that small deviations from perfect hydrostatic balance hardly matters.

These thoughts are of course a matter for the flume laboratory, field tests and calibrations.

Classical Stokesian Theory treats of a rigid geometrically-ideal body gravitating through a viscous fluid, with no turbulence and only laminar skin flow impinging upon the interface.

First of all, we must supply μ, the Dynamic Viscosity of Water (in Pascal.Seconds). μ is sensitive to temperature. The Mean English Surface Water Temperature[3], T_{ang}, in Degrees Celsius, is 10.97.

The equation Water Temperature, T_w, is accordingly 284.12 Degrees Kelvin.

The relevant μ_w is approximated to four figures by:-

$$\mu_w = 0.00002414 \times 10^{\frac{247.8}{T_w - 140}}$$

Equation 55

This gives μ_w = 0.001265137606259 Pascal.Seconds at 10.97°C.

As a working assumption let us say that the finished float is one milligram too heavy. I.e. ΔMass = 0.000001 kg.

Then the Actual Float Mass, M_{act}, is:-

$$M_{act} = M_{SP} + M_b + M_{ca} + \Delta Mass$$

Equation 56

Where V_D is the Displaced Volume it follows that the Actual Mean Float Density, ρ_{act}, is:-

$$\rho_{act} = \frac{M_{act}}{V_D}$$

Equation 57

and that system Differential Density, $\Delta\rho$, is:-

$$\Delta\rho = \rho_{act} - \rho_w$$

Equation 58

The (settling or emergent) Terminal Velocity, v_t, is yielded by:-

$$V_t = \frac{2}{9} \cdot \frac{\Delta\rho}{\mu_w} \cdot gr_p^{\ 2}$$

Equation 59

neglecting the unbalanced effects of the painted label.

For our ping-pong ball float with Δmass = 0.000001, v_t = 0.02040376730695, a sinking at over two centimeters per second that would bring our float to a stream bed at 50 centimeters in 25 seconds.

Taking the reciprocal approach, let is declare that our design requirement is for a Required Minimum Fall or Rise Velocity, $Vreq_t$, of 0.001, i.e. 1mm/second.

Some transposition and substitution furnishes an expression for the required ΔMass tolerance as:-

$$\Delta Mass = \left[V_D \left(\frac{9}{2} \cdot \frac{Vreq_t}{gr_p{}^2} \cdot \mu_w \right) + M_b \right] - \left(M_{SP} + M_b + M_{ca} \right)$$

Equation 60

From which Δmass results as 0.000000049010557, a tolerance of ±49 micrograms.

The tolerances required are so fine that effervescent water or some other fluid with a fugacious component, or a fugacious fluid, may well suggest itself as a ballast for hand-made floats.

PART FOUR
REFERENCES

1 **Stokesian Velocity**

 http://en.wikipedia.org/wiki/Stoke%27s_Law

2 **Cubic Function**

 http://en.wikipedia.org/wiki/Cubic_equation

3 **English Water Temperature**

"Changing water temperatures:
 a Surface Water Archive for England and Wales"
HG Orr et al
pp 8
http://nora.nerc.ac.uk/12050/
British Hydrological Society Third International Symposium
Newcastle 2010
http://ceg.ncl.ac.uk/bhs2010/

APPENDIX ONE
NOTATION AND DATA

Component	Mass	Density	Volume	Radius	Thickness	Depth
Contained Air	M_{ca}	ρ_{ca}	V_{ca}	r_{ca}	t_{ca}	d_{ca}
Contained Fluid	M_f	ρ_f	V_f	r_f	t_f	d_f
Contained Granules	M_g	ρ_g	V_g	r_g	t_g	d_g
Contained Ballast	M_b	ρ_b	V_b	r_b	t_b	d_b
Shell	M_s	ρ_s	V_s	r_s	t_s	d_s
Plug	M_u	ρ_u	V_u	r_u	t_u	d_u
Paint Layer	M_p	ρ_p	V_p	r_p	t_p	d_p
Label Layer	M_l	ρ_l	V_l	r_l	t_l	d_l
Plug Material	M_e	ρ_e	V_e	r_e	t_e	d_e
Stream Water	M_w	ρ_w	V_w	r_w	t_w	d_w

Table One
Materials Notation

Description	Symbol	Description	Symbol
Force exerted Upward	F_{up}	Laminar Frictional Force (Stokesian)	F_f
Force exerted Downward	F_{down}	Dynamic Viscosity	μ
Displaced Fluid Mass	M_D	Object Radius	R
Mass of Assembled Float	M_{float}	Terminal Velocity	v_t
Displaced Fluid Volume	V_D	Composite Density of Float	ρ_{float}
The local Acceleration Due to Gravity	g	Density of Water	ρ_w
Label Text Area	LTA	The Square Root of Minus One	i
Under Paint Area	UPA	Coefficient Zero	k_0
Fractional Label Area	f_i	Coefficient One	k_1
Shell Thickness	t_s	Coefficient Two	k_2
Available Chamber Thickness	V_C	Coefficient Three	k_3
		Discriminant	Δ
Plug Radius	r	Areal Mass Density Cofactor	CF
Plug Radius	a_1	The Real Part of Q	q
Plug Radius	a_2	Imaginary Mass Density Cofactor	Q
Inner Cap Height	h_1	Real Mass Density Cofactor	R
Outer Cap Height	h_2	Cube Root of Complex Mass Density Cofactor	C
Central Plug Depth	d_u	First Root of Cubic Equation	x_1
Volume of First Cap	V_{cap1}	Second Root of Cubic Equation	x_2
Volume of Second Cap	V_{cap2}	Third Root of Cubic Equation	x_3
Volume of Cylinder	V_{cyl}		
		Central Depth of Fluid Ballast	d_b
Plug Mass Differential	$M_{\Delta u}$	Volume of Ballast	V_b
Chamber Content Target Mass	M_T	Mass of Ballast	M_b
Mass of Air in Full Unballasted Chamber	M_F	Volume of Contained Air	V_{ca}
Whole-Chamber Mass Differential	ΔM_C	Mass of Contained Air	M_{ca}
Infill Differential Density	$\Delta \rho_\alpha$		
		Enclosed Chamber Volume (1)	$V_b + V_{ca}$
Density of Nickel	ρ_{Ni}	Enclosed Chamber Volume (2)	$(4/3)\pi r_{ca}^3$
Ballast Sphere Diameter	d_{ball}		
Ballast Sphere Volume	V_{ball}	Enclosed Chamber Mass	$M_b + M_{ca}$
Ballast Sphere Mass	M_{ball}	Chamber Content Target Mass	M_T
Number of Required Ballast Spheres	N_{balls}		
Float Ballast Cost	Cost	Water Temperature in Degrees °Kelvin	T_w
		Mean English Surface water Temperature in °C	T_{ang}
Density of Stainless Steel	ρ_{ss}	Dynamic Viscosity of Water at T_w °K	μ_w
Diameter of Wire	d_{wire}	Differential Mass	$\Delta Mass$
Radius of Wire	r_{wire}	Differential Density	$\Delta \rho$
Volume of Wire	V_{wire}	Actual Float Mass	M_{act}
Length of Wire	L_{wire}	Actual Float Mean Density	ρ_{act}
		Terminal Settling Velocity	v_t
		Required Terminal Settling Velocity (limit)	$vreq_t$

Table Two
Classified General Notation

Air density is at STP
Water Density is at 20 degrees C
Ping-pong balls are assumed to be made of cellulose nitrate
and to be the International sporting standard 40mm diameter
of 2.7 grams weight with a Coefficient of Restitution of 0.4
The density of epoxy putty is based upon a 2 US ounce cylindrical stick
of "Quiksteel" as a 4-inch cylinder of 1.0625 inches diameter
(http://www.dual-star.com/index2/Service/QuikSteel%20Epoxy%20Putty.htm)
The density of dry epoxy paint is computed from knowledge that
ordinary Chinese epoxy ship paint is 1450kg/cu.m wet
(Compartment Epoxy Paint (GLC-EPTANK 366))
and that Akzo-Nobel Intershield 803Plus EGA-903Red
Abrasion Resistant Binary Epoxy Paint is designed to leave a
125 micron coat at 5sq.m/liter for 71% dry solids (computed value is 62.5%)
(http://www.international-marine.com/PDS/4313+M+eng+A4.pdf)
Ballast Wire and Sphere Materials:-
(a) Nickel Wire 99% 0.5mm dia Annealed NI005151
5 meters = £80
(http:/www.goodfellow.com/E/Nickel-Wire.html)
(b) Nickel Sphere 99% 0.76mm dia NI006811
50 spheres = £66.50
(http:/www.goodfellow.com/E/Nickel-Sphere.html)
(c) Nickel Wire 200 99% 0.39mm dia Annealed and Cold-drawn in Spool
100 meters = £45
(http://www.alloyshop.com/products/039mm-dia-nickel-200)
(d) Nickel Wire 0.5mm dia 25 SWG Ref.: NI0500-050
~28 meters = 50grams = £11.45exVAT%
(http://wires.co.uk/acatalog/ni_bare.html)
(e) Stainless Steel 0.5mm dia Bare Ref.: SS0500-050
50grams = £4exVAT%
(http://wires.co.uk/acatalog/)

Table Three
Notes upon Materials and Components

Substance	Density ρ (kg/m^3)	Error Bound (+/- Kgs)
Air	1.2041	
Mercury	13546.0000	
Lead	11340.0000	
Tin	7365.0000	
Nickel	8908.0000	
Titanium	4506.0000	
Stainless Steel	7900.0000	
Glass	2579.0000	
Water	998.0000	
Cellulose Nitrate	1375.0000	25
Epoxy Putty	975.5889	
Dry Paint	1721.2000	

Table Four
Densities of Materials

APPENDIX TWO
MERCURY:AIR BALLAST COUPLE
EXCEL® SPREADSHEET
DATA

Component	Mass	Density	Volume	Radius	Thickness	Depth
Contained Air	0.000037985363739	1.20410000	0.000031546685275	0.019601457920705	t_{ca}	d_{ca}
Contained Fluid	M_f	ρ_f	V_f	r_f	t_f	d_f
Contained Granules	M_g	ρ_g	V_g	r_g	t_g	d_g
Contained Ballast	0.030215095969246	13546.00000000	0.000002230554848	r_b	t_b	0.006373821991182
Shell	0.002700000000000	1375.00000000	0.000001963636364	0.020000000000000	0.000398542079295	d_s
Plug	0.000001222272743	ρ_u	0.000000001252856	0.001000000000000	t_u	d_u
Paint Layer	0.001088235073201	1721.20000000	0.000000632253703	0.020125000000000	0.000125000000000	d_p
Label Layer	0.000086060000000	ρ_l	0.000000050000000	r_l	t_l	d_l
Plug Material	M_e	975.58890717	V_e	r_e	t_e	d_e
Stream Water	0.034124190190375	998.00000000	0.000034192575341	r_w	t_w	d_w
Shell, Plug and Paint	0.003873794668483					
Displaced Material	0.034124190190375					
Resolved Contained Air	0.000035299552647		0.000029316130427			
Resolved Contained Ballast	0.030215095969246		0.000002230554848			

Table Five
Mercury:Air
Buoyancy Model Transput Matrix

Label Text Area	0.000400000000000
Under Paint Area	0.005089576448356
Fractional Label Area	0.078592001526803
Plug Radius	0.001000000000000
Inner Cap Height	0.000025524925898
Outer Cap Height	0.000025015644562
Central Plug Depth	0.000399051360631
Volume of First Cap	0.000000000040103
Volume of Second Cap	0.000000000039303
Volume of Cylinder	0.000000001253657
Plug Mass Differential	-0.000000500404718
Chamber Content Target Mass	0.030250395521892
Mass of Air in Full Unballasted Chamber	0.000037985363739
Whole-Chamber Mass Differential	0.427293413366735
Infill Differential Density	13544.795900000000000

Table Six
Mercury:Air
Ancillary Results

Coefficient Zero	-0.030212
Coefficient One	0.000000
Coefficient Two	834.085771
Coefficient Three	-14184.077098
Discriminant	65167402.823077
CF	695699.073615
q	594974252.155915
Q	594974252.155915i
R	996429253.294240
R+Q	996429253.29424+594974252.155915i
(R+Q)^(1/3)	1034.01097118916+187.549334228117i
C	820.695051706166+148.858005224824i
1+I*3^(1/2)	1+1.73205080756888i
1-I*3^(1/2)	1-1.73205080756888i
C*(1+I*3^(1/2))	562.865423543417+1570.34353230027i
CF*(1-I*3^(1/2))/C	562.865423543422-1570.34353230029i
C*(1-I*3^(1/2))	1078.52467986892-1272.62752185062i
CF*(1+I*3^(1/2))/C	1078.52467986893+1272.62752185064i
x1	0.058174995747169
x2	0.006373821991182
x3	-0.005744443976236
$\mathbf{V_b}$	0.000002230554848
$\mathbf{M_b}$	0.030215095969246
$\mathbf{V_{ca}}$	0.000029316130427
$\mathbf{M_{ca}}$	0.000035299552647
$\mathbf{V_b+V_{ca}}$	0.000031546685275
$\mathbf{(4/3)pr_{ca}{}^3}$	0.000031546685275
$\mathbf{M_b+M_{ca}}$	0.030250395521893
MT	0.030250395521892

Table Seven
Mercury:Air
Model Cubic Equations Solutions List

APPENDIX THREE
MERCURY:AIR BALLAST COUPLE
EXCEL® SPREADSHEET
FORMULAE

	A	B	C
1	Component	Mass	Density
2	Contained Air	=C2*D2	1.2041
3	Contained Fluid	M_f	ρ_f
4	Contained Granules	M_g	ρ_g
5	Contained Ballast	='CUBIC SOLUTIONS LIST'!B29	13546
6	Shell	0.0027	1375
7	Plug	=C10*D7	ρ_u
8	Paint Layer	=C8*D8	1721.2
9	Label Layer	=C8*D9	ρ_l
10	Plug Material	M_e	975.58890717
11			
12	Stream Water	=B15	998
13			
14	Shell, Plug and Paint	=B6+'ANCILLARY RESULTS'!B13+B8+B9	
15	Displaced Material	=C12*((4/3)*PI()*E8^3+D9)	
16			
17	Resolved Contained Air	='CUBIC SOLUTIONS LIST'!B32	
18	Resolved Contained Ballast	='CUBIC SOLUTIONS LIST'!B29	

Table Eight
Buoyancy Model Transput Matrix
Formulae Page 1

	D	E	F
1	Volume	Radius	Thickness
2	=(4/3)*PI()*E2^3	=(E6^3-(3*D6)/(4*PI()))^(1/3)	t_{ce}
3	V_f	r_f	t_f
4	V_g	r_g	t_g
5	='CUBIC SOLUTIONS LIST'IB28	r_b	t_b
6	=B6/C6	0.02	=E6-E2
7	='ANCILLARY RESULTS'IB11+'ANCILLARY RESULTS'IB10-'ANCILLARY RESULTS'IB9	0.001	t_u
8	=(4/3)*PI()*F8*(3*E6*E8+F8^2)	=E6+F8	0.000125
9	=F8*'ANCILLARY RESULTS'IB1	r_i	t_i
10	V_e	r_e	t_e
11			
12	=(4/3)*PI()*E8^3+D9	r_w	t_w
13			
14			
15			
16			
17	='CUBIC SOLUTIONS LIST'IB31		
18	='CUBIC SOLUTIONS LIST'IB28		

Table Nine
Buoyancy Model Transput Matrix
Formulae Page 2

	G
1	Depth
2	d_{ca}
3	d_f
4	d_g
5	='CUBIC SOLUTIONS LIST'!B24
6	d_s
7	d_u
8	d_p
9	d_l
10	d_e
11	
12	d_w
13	
14	
15	
16	
17	
18	

VOLFLOATmercuryairFORMULAE.xls   Prepared by James R Warren   13:16 07/12/2011

Table Ten
Buoyancy Model Transput Matrix
Formulae Page 3

	A	B
1	Label Text Area	0.0004
2	Under Paint Area	=4*PI()*'BUOYANCY MODEL TRANSPUT MATRIX'!E8^2
3	Fractional Label Area	=B1/B2
4		
5	Plug Radius	='BUOYANCY MODEL TRANSPUT MATRIX'!E7
6	Inner Cap Height	='BUOYANCY MODEL TRANSPUT MATRIX'!E2-SQRT('BUOYANCY MODEL TRANSPUT MATRIX'!E2^2-B5^2)
7	Outer Cap Height	='BUOYANCY MODEL TRANSPUT MATRIX'!E6-SQRT('BUOYANCY MODEL TRANSPUT MATRIX'!E6^2-B5^2)
8	Central Plug Depth	='BUOYANCY MODEL TRANSPUT MATRIX'!F6+B6-B7
9	Volume of First Cap	=(PI()*B6^2/3)*(3*'BUOYANCY MODEL TRANSPUT MATRIX'!E2-B6)
10	Volume of Second Cap	=(PI()*B7^2/3)*(3*'BUOYANCY MODEL TRANSPUT MATRIX'!E6-B7)
11	Volume of Cylinder	=PI()*B5^2*B8
12		
13	Plug Mass Differential	='BUOYANCY MODEL TRANSPUT MATRIX'!B7-'BUOYANCY MODEL TRANSPUT MATRIX'!D7*'BUOYANCY MODEL TRANSPUT MATRIX'!C6
14	Chamber Content Target Mass	='BUOYANCY MODEL TRANSPUT MATRIX'!B15-'BUOYANCY MODEL TRANSPUT MATRIX'!B14
15		
16	Mass of Air in Full Unballasted Chamber	='BUOYANCY MODEL TRANSPUT MATRIX'!B2
17	Whole-Chamber Mass Differential	='BUOYANCY MODEL TRANSPUT MATRIX'!D2*('BUOYANCY MODEL TRANSPUT MATRIX'!C5-'BUOYANCY MODEL TRANSPUT MATRIX'!C2)
18	Infill Differential Density	='BUOYANCY MODEL TRANSPUT MATRIX'!C5-'BUOYANCY MODEL TRANSPUT MATRIX'!C2

Table Eleven
Ancillary Results
Formulae

	A	B
1	Coefficient Zero	='BUOYANCY MODEL TRANSPUT MATRIX'!B2-'ANCILLARY RESULTS'!B14
2	Coefficient One	0
3	Coefficient Two	=PI()*'BUOYANCY MODEL TRANSPUT MATRIX'!E2*('BUOYANCY MODEL TRANSPUT MATRIX'!C5-'BUOYANCY MODEL TRANSPUT MATRIX'!C2)
4	Coefficient Three	=(PI()/3)*('BUOYANCY MODEL TRANSPUT MATRIX'!C2-'BUOYANCY MODEL TRANSPUT MATRIX'!C5)
5		
6	Discriminant	=18*B4*B3*B2*B1-4*B3^3*B1+B3^2*B2^2-4*B4*B2^3-27*B4^2*B1^2
7	CF	=B3^2
8		
9	q	=3*PI()^2*'ANCILLARY RESULTS'!B18^2*SQRT('ANCILLARY RESULTS'!B17+'ANCILLARY RESULTS'!B16-'ANCILLARY RESULTS'!B14)*SQRT('ANCILLARY RESULTS'!B14-'ANCILLARY RESULTS'!B16)
10	Q	=COMPLEX(0,B9)
11	R	=3*PI()^2*'ANCILLARY RESULTS'!B18^2*(0.5*'ANCILLARY RESULTS'!B17+'ANCILLARY RESULTS'!B16-'ANCILLARY RESULTS'!B14)
12	R+Q	=COMPLEX(B11,B9)
13	(R+Q)^(1/3)	=IMPOWER(B12,1/3)
14	C	=COMPLEX(0.5^(1/3)*IMREAL(B13),0.5^(1/3)*IMAGINARY(B13))
15		
16	1+I*3^(1/2)	=COMPLEX(1,3^(1/2))
17	1-I*3^(1/2)	=IMCONJUGATE(B16)
18	C*(1+I*3^(1/2))	=IMPRODUCT(B14,B16)
19	CF*(1-I*3^(1/2))/C	=IMDIV(IMPRODUCT(COMPLEX(B7,0),IMCONJUGATE(B16)),B14)
20	C*(1-I*3^(1/2))	=IMPRODUCT(B14,B17)
21	CF*(1+I*3^(1/2))/C	=IMDIV(IMPRODUCT(COMPLEX(B7,0),IMCONJUGATE(B17)),B14)
22		
23	x1	=IMREAL(IMPRODUCT(COMPLEX(-(1/(3*B4)),0),IMSUM(COMPLEX(B3,0),IMSUM(B14,IMDIV(COMPLEX(B7,0),B14)))))
24	x2	='BUOYANCY MODEL TRANSPUT MATRIX'!E2-(1/(2*PI()*'ANCILLARY RESULTS'!B18))*IMABS(IMSUM(B18,B19))
25	x3	='BUOYANCY MODEL TRANSPUT MATRIX'!E2-(1/(2*PI()*'ANCILLARY RESULTS'!B18))*IMABS(IMSUM(B20,B21))
26		
27		
28	V_b	=(PI()*B24^2/3)*(3*'BUOYANCY MODEL TRANSPUT MATRIX'!E2-B24)
29	M_b	='BUOYANCY MODEL TRANSPUT MATRIX'!C5*B28
30		
31	V_{ca}	=(4/3)*PI()*'BUOYANCY MODEL TRANSPUT MATRIX'!E2^3-(PI()*B24^2/3)*(3*'BUOYANCY MODEL TRANSPUT MATRIX'!E2-B24)
32	M_{ca}	='BUOYANCY MODEL TRANSPUT MATRIX'!C2*B31
33		
34	V_b+V_{ca}	=B28+B31
35	$(4/3)pr_{ca}^3$	=(4/3)*PI()*'BUOYANCY MODEL TRANSPUT MATRIX'!E2^3
36		
37	M_b+M_{ca}	=B29+B32
38	MT	='ANCILLARY RESULTS'!B14

Table Twelve
Cubic Solutions List
Formulae Page 1

Component	Mass	Density	Volume	Radius	Thickness	Depth
Contained Air	0.031483591904106	998.00000000	0.000031546685275	0.019601457920705	t_{ca}	d_{ca}
Contained Fluid	M_f	ρ_f	V_f	r_f	t_f	d_f
Contained Granules	M_g	ρ_g	V_g	r_g	t_g	d_g
Contained Ballast	0.000001489664799	1.20410000	0.000001237160368	r_b	t_b	0.004671632604740
Shell	0.002700000000000	1375.00000000	0.000001963636364	0.020000000000000	0.000398542079295	d_s
Plug	0.000001222272743	ρ_u	0.000000001252856	0.001000000000000	t_u	d_u
Paint Layer	0.001088235073201	1721.20000000	0.000000632253703	0.020125000000000	0.000125000000000	d_p
Label Layer	0.000086060000000	ρ_l	0.000000050000000	r_l	t_l	d_l
Plug Material	M_e	975.58890717	V_e	r_e	t_e	d_e
Stream Water	0.034124190190375	998.00000000	0.000034192575341	r_w	t_w	d_w
Shell, Plug and Paint	0.003873794668483					
Displaced Material	0.034124190190375					
Resolved Contained Air	0.030248905857093		0.000030309524907			
Resolved Contained Ballast	0.000001489664799		0.000001237160368			

Table Thirteen
Water:Air
Buoyancy Model Transput Matrix

Label Text Area	0.000400000000000
Under Paint Area	0.005089576448356
Fractional Label Area	0.078592001526803
Plug Radius	0.001000000000000
Inner Cap Height	0.000025524925898
Outer Cap Height	0.000025015644562
Central Plug Depth	0.000399051360631
Volume of First Cap	0.000000000040103
Volume of Second Cap	0.000000000039303
Volume of Cylinder	0.00000001253657
Plug Mass Differential	-0.000000500404718
Chamber Content Target Mass	0.030250395521892
Mass of Air in Full Unballasted Chamber	0.031483591904106
Whole-Chamber Mass Differential	-0.031445606540366
Infill Differential Density	-996.795900000000000

Table Fourteen
Water:Air
Ancillary Results

	$\mathfrak{Re}$	$\mathfrak{Im}$
Coefficient Zero	0.001233196382213	
Coefficient One	0.000000000000000	
Coefficient Two	-61.382488378320900	
Coefficient Three	1043.842225522810000	
Discriminant	1096.102327718180000	
CF	3767.809879514700000	
U	-426275.221792411000000	
Q	0.000000000010995	179573.518877097000000
R	-426275.221792411000000	
C	37.464630758747800	48.623156228544000
CF/C	37.464630758747700	-48.623156228543900
1+I*3^(1/2)	1.000000000000000	1.732050807568880
1-I*3^(1/2)	1.000000000000000	-1.732050807568880
C*(1+I*3^(1/2))	-46.753146253449700	113.513800189503000
CF*(1-I*3^(1/2))/C	-46.753146253449500	-113.513800189503000
C*(1-I*3^(1/2))	121.682407770945000	-16.267487732415000
CF*(1+I*3^(1/2))/C	121.682407770945000	16.267487732415000
Term11	0.019601457920705	
Term12	-0.011963695228617	-0.015526981325871
Term13	-0.011963695228617	0.015526981325871
Term21	0.019601457920705	
Term22	-0.007464912657982	0.018124354654052
Term23	-0.007464912657982	-0.018124354654052
Term31	0.019601457920705	
Term32	0.019428607886599	-0.002597373328181
Term33	0.019428607886599	0.002597373328181
x1	-0.004325932536529	
x2	0.004671632604740	
x3	0.058458673693903	
V_b	0.000001237160368	
M_b	0.000001489664799	
V_{ca}	0.000030309524907	
M_{ca}	0.030248905857093	
V_b+V_{ca}	0.000031546685275	
$(4/3)pr_{ca}^{3}$	0.000031546685275	
M_b+M_{ca}	0.030250395521892	
M_T	0.030250395521892	

Table Fifteen
Water:Air
Model Cubic Equations Solutions List

APPENDIX FIVE
WATER:AIR BALLAST COUPLE
EXCEL® SPREADSHEET
FORMULAE

	A	B	C
1	Component	Mass	Density
2	Contained Air	=C2*D2	998
3	Contained Fluid	M_f	ρ_f
4	Contained Granules	M_g	ρ_g
5	Contained Ballast	='CUBIC SOLUTIONS LIST'!B37	1.2041
6	Shell	0.0027	1375
7	Plug	=C10*D7	ρ_u
8	Paint Layer	=C8*D8	1721.2
9	Label Layer	=C8*D9	ρ_l
10	Plug Material	M_e	975.58890717
11			
12	Stream Water	=B15	998
13			
14	Shell, Plug and Paint	=B6+'ANCILLARY RESULTS'!B13+B8+B9	
15	Displaced Material	=C12*((4/3)*PI()*E8^3+D9)	
16			
17	Resolved Contained Air	='CUBIC SOLUTIONS LIST'!B40	
18	Resolved Contained Ballast	='CUBIC SOLUTIONS LIST'!B37	

Table Sixteen
Buoyancy Model Transput Matrix
Formulae Page 1

BUOYANCY MODEL TRANSPUT MATRIX

	D	E	F
1	Volume	Radius	Thickness
2	=(4/3)*PI()*E2^3	=(E6^3-(3*D6)/(4*PI()))^(1/3)	t_{os}
3	V_f	r_f	t_f
4	V_g	r_g	t_g
5	='CUBIC SOLUTIONS LIST'!B36	r_b	t_b
6	=B6/C6	0.02	=E6-E2
7	='ANCILLARY RESULTS'!B11+'ANCILLARY RESULTS'!B10-'ANCILLARY RESULTS'!B9	0.001	t_u
8	=(4/3)*PI()*F8*(3*E6*E8+F8^2)	=E6+F8	0.000125
9	=F8*'ANCILLARY RESULTS'!B1	r_i	t_i
10	V_e	r_e	t_e
11			
12	=(4/3)*PI()*E8^3+D9	r_w	t_w
13			
14			
15			
16			
17	='CUBIC SOLUTIONS LIST'!B39		
18	='CUBIC SOLUTIONS LIST'!B36		

Table Seventeen
Buoyancy Model Transput Matrix
Formulae Page 2

	G
1	**Depth**
2	d_{ca}
3	d_f
4	d_g
5	=TEXT('CUBIC SOLUTIONS LIST'!B33,"0.000000000000000")
6	d_s
7	d_u
8	d_p
9	d_l
10	d_e
11	
12	d_w
13	
14	
15	
16	
17	
18	

Table Eighteen
Buoyancy Model Transput Matrix
Formulae Page 3

Label Text Area	0.0004
Under Paint Area	=4*PI()*'BUOYANCY MODEL TRANSPUT MATRIX'!E8^2
Fractional Label Area	=B1/B2
Plug Radius	='BUOYANCY MODEL TRANSPUT MATRIX'!E7
Inner Cap Height	='BUOYANCY MODEL TRANSPUT MATRIX'!E2-SQRT('BUOYANCY MODEL TRANSPUT MATRIX'!E2^2-B5^2)
Outer Cap Height	='BUOYANCY MODEL TRANSPUT MATRIX'!E6-SQRT('BUOYANCY MODEL TRANSPUT MATRIX'!E6^2-B5^2)
Central Plug Depth	='BUOYANCY MODEL TRANSPUT MATRIX'!F6+B6-B7
Volume of First Cap	=(PI()*B6^2/3)*(3*'BUOYANCY MODEL TRANSPUT MATRIX'!E2-B6)
Volume of Second Cap	=(PI()*B7^2/3)*(3*'BUOYANCY MODEL TRANSPUT MATRIX'!E6-B7)
Volume of Cylinder	=PI()*B5^2*B8
Plug Mass Differential	='BUOYANCY MODEL TRANSPUT MATRIX'!B7-'BUOYANCY MODEL TRANSPUT MATRIX'!D7*'BUOYANCY MODEL TRANSPUT MATRIX'!C6
Chamber Content Target Mass	='BUOYANCY MODEL TRANSPUT MATRIX'!B15-'BUOYANCY MODEL TRANSPUT MATRIX'!B14
Mass of Air in Full Unballasted Chamber	='BUOYANCY MODEL TRANSPUT MATRIX'!B2
Whole-Chamber Mass Differential	='BUOYANCY MODEL TRANSPUT MATRIX'!D2*('BUOYANCY MODEL TRANSPUT MATRIX'!C5-'BUOYANCY MODEL TRANSPUT MATRIX'!C2)
Infill Differential Density	='BUOYANCY MODEL TRANSPUT MATRIX'!C5-'BUOYANCY MODEL TRANSPUT MATRIX'!C2

Table Nineteen
Ancillary Results
Formulae Page 1

Coefficient Zero	='BUOYANCY MODEL TRANSPUT MATRIX'!B2-'ANCILLARY RESULTS'!B14
Coefficient One	0
Coefficient Two	=PI()*'BUOYANCY MODEL TRANSPUT MATRIX'!E2*('BUOYANCY MODEL TRANSPUT MATRIX'!C5-'BUOYANCY MODEL TRANSPUT MATRIX'!C2)
Coefficient Three	=(PI()/3)*('BUOYANCY MODEL TRANSPUT MATRIX'!C2-'BUOYANCY MODEL TRANSPUT MATRIX'!C5)
Discriminant	=18*B5*B4*B3*B2-4*B4^3*B2+B4^2*B3^2-4*B5*B3^3-27*B5^2*B2^2
CF	=B4^2
U	=2*B4^3-9*B5*B4*B3+27*B5^2*B2
Q	=IMSQRT(B9^2-4*B8^3)
R	=3*PI()^2*'ANCILLARY RESULTS'!B18^2*(0.5*'ANCILLARY RESULTS'!B17+'ANCILLARY RESULTS'!B16-'ANCILLARY RESULTS'!B14)
C	=IMPOWER(IMPRODUCT(0.5,IMSUM(B10,B11)),(1/3))
CF/C	=IMDIV(COMPLEX(B8,0),B12)
1+I*3^(1/2)	=COMPLEX(1,3^(1/2))
1-I*3^(1/2)	=IMCONJUGATE(B15)
C*(1+I*3^(1/2))	=IMPRODUCT(B12,B15)
CF*(1-I*3^(1/2))/C	=IMDIV(IMPRODUCT(COMPLEX(B8,0),IMCONJUGATE(B15)),B12)
C*(1-I*3^(1/2))	=IMPRODUCT(B12,B16)
CF*(1+I*3^(1/2))/C	=IMDIV(IMPRODUCT(COMPLEX(B8,0),IMCONJUGATE(B16)),B12)
Term11	=-B4/(3*B5)
Term12	=IMDIV(B12,-(3*B5))
Term13	=IMDIV(B13,-3*B5)
Term21	=B22
Term22	=IMDIV(B17,(6*B5))
Term23	=IMDIV(B18,(6*B5))
Term31	=B22
Term32	=IMDIV(B19,(6*B5))
Term33	=IMDIV(B20,(6*B5))
x1	=IMSUM(B22,B23,B24)
x2	=IMSUM(B25,B26,B27)
x3	=IMSUM(B28,B29,B30)
V_b	=(PI()*B33^2/3)*(3*'BUOYANCY MODEL TRANSPUT MATRIX'!E2-B33)
M_b	='BUOYANCY MODEL TRANSPUT MATRIX'!C5*B36
V_{ca}	=(4/3)*PI()*'BUOYANCY MODEL TRANSPUT MATRIX'!E2^3-(PI()*B33^2/3)*(3*'BUOYANCY MODEL TRANSPUT MATRIX'!E2-B33)
M_{ca}	='BUOYANCY MODEL TRANSPUT MATRIX'!C2*B39
V_b+V_{ca}	=B36+B39
$(4/3)pr_{ca}{}^3$	=(4/3)*PI()*'BUOYANCY MODEL TRANSPUT MATRIX'!E2^3
M_b+M_{ca}	=B37+B40
M_T	='ANCILLARY RESULTS'!B14

Table Twenty
Cubic Solutions List
Formulae Page 1

Aspects of the Design of Neutrally-Buoyant
Solid Ballasted Tachometric Floats

by
James R Warren BSc MSc PhD PGCE

PART ONE
SOME BALLAST MEDIA

Several forms and substances of ballast might suggest themselves.
A non-exhaustive list of possibilities might include:-
(a)   Mercury
(b)   Still or Sparkling Water
(c)   Gases liquid near 10°C
(d)   Spherical Shot
(e)   Wire
(f)   Barium Powders
Or of course some inert mixture of these or other fillers.

PART TWO
SOME SOLID BALLAST MEDIA

<u>Shot</u>

Problems with shot include the fact of packing vagaries generally, the fact that packing does not complete any design-cap evacuation, and of course discretisation.
Consider commercial NI006811 99% Nickel 0.76mm diameter spheres at a nominal cost of £66.50 per 50.
The Density of Nickel, ρ_{Ni} = 8908 and Ball Diameter, d_{ball}, is 0.00076.
The Mass of One Nickel Ball, M_{ball}, is:-

$$M_{ball} = \rho_{Ni}.V_{ball} = \rho_{Ni}.\frac{4}{3}\pi\left(\frac{d_{ball}}{2}\right)^3$$

Equation 1

whilst the Required Number of Balls, N_{balls}, is:-

$$N_{balls} = \frac{M_b}{M_{ball}}$$

Equation 2

The computed requirement is 14757.21383974628 Nickel Spheres at a Cost of £19,627.09 per float.

Because M_{ball} = 0.000002047479714 it is clear that neither 14757 nor 14758 inserted nickel spheres would achieve hydrostatic stability.

<u>Wire</u>

Problems with wire include insertion issues, and winding and winding land design, if winding is attempted. In contrast to shot if wire is inserted into the chamber it over-completes cap evacuation.

Interested researchers or designers could look into the implications of loxodromic winding within or outwith the float.

Consider commercial Stainless Steel 0.5mm diameter Bare Wire SS0500-050 at a nominal Cost of £4.80 for 50 grams.

The Density of Stainless Steel, ρ_{ss}, is taken to be 7900 and Wire Diameter, d_{wire}, is 0.0005 meters. Clearly, Wire Radius, half the diameter, is r_{wire} = 0.00025.

The Required Volume of Wire, V_{wire}, is given by:-

$$V_{wire} = \frac{M_b}{\rho_{ss}}$$

Equation 3

From which the Required Wire Length, L_{wire}, is computed by:-

$$L_{wire} = \frac{V_{wire}}{\pi r_{wire}^{2}}$$

Equation 4

Since M_b = 0.030215095969245 kg; V_{wire} = 0.000003824695692 m^3 and L_{wire} = 19.47901520810689 meters, the Cost per float of the wire ballast is £2.42 ex VAT.

PART THREE
REFERENCES

1   **Cubic Function**

http://en.wikipedia.org/wiki/Cubic_equation

Component	Mass	Density	Volume	Radius	Thickness	Depth
Contained Air	M_{ca}	ρ_{ca}	V_{ca}	r_{ca}	t_{ca}	d_{ca}
Contained Fluid	M_f	ρ_f	V_f	r_f	t_f	d_f
Contained Granules	M_g	ρ_g	V_g	r_g	t_g	d_g
Contained Ballast	M_b	ρ_b	V_b	r_b	t_b	d_b
Shell	M_s	ρ_s	V_s	r_s	t_s	d_s
Plug	M_u	ρ_u	V_u	r_u	t_u	d_u
Paint Layer	M_p	ρ_p	V_p	r_p	t_p	d_p
Label Layer	M_l	ρ_l	V_l	r_l	t_l	d_l
Plug Material	M_e	ρ_e	V_e	r_e	t_e	d_e
Stream Water	M_w	ρ_w	V_w	r_w	t_w	d_w

**Table One
Materials Notation**

Force exerted Upward	F_{up}	Laminar Frictional Force (Stokesian)	F_f
Force exerted Downward	F_{down}	Dynamic Viscosity	μ
Displaced Fluid Mass	M_D	Object Radius	R
Mass of Assembled Float	M_{float}	Terminal Velocity	v_t
Displaced Fluid Volume	V_D	Composite Density of Float	ρ_{float}
The local Acceleration Due to Gravity	g	Density of Water	ρ_w
Label Text Area	LTA	The Square Root of Minus One	i
Under Paint Area	UPA	Coefficient Zero	k_0
Fractional Label Area	f_l	Coefficient One	k_1
Shell Thickness	t_s	Coefficient Two	k_2
Available Chamber Thickness	V_C	Coefficient Three	k_3
		Discriminant	Δ
Plug Radius	r	Areal Mass Density Cofactor	CF
Plug Radius	a_1	The Real Part of Q	q
Plug Radius	a_2	Imaginary Mass Density Cofactor	Q
Inner Cap Height	h_1	Real Mass Density Cofactor	R
Outer Cap Height	h_2	Cube Root of Complex Mass Density Cofactor	C
Central Plug Depth	d_u	First Root of Cubic Equation	x_1
Volume of First Cap	V_{cap1}	Second Root of Cubic Equation	x_2
Volume of Second Cap	V_{cap2}	Third Root of Cubic Equation	x_3
Volume of Cylinder	V_{cyl}		
		Central Depth of Fluid Ballast	d_b
Plug Mass Differential	$M_{\Delta u}$	Volume of Ballast	V_b
Chamber Content Target Mass	M_T	Mass of Ballast	M_b
Mass of Air in Full Unballasted Chamber	M_F	Volume of Contained Air	V_{ca}
Whole-Chamber Mass Differential	ΔM_C	Mass of Contained Air	M_{ca}
Infill Differential Density	$\Delta\rho_\alpha$		
		Enclosed Chamber Volume (1)	V_b+V_{ca}
Density of Nickel	ρ_{Ni}	Enclosed Chamber Volume (2)	$(4/3)\pi r_{ca}^{3}$
Ballast Sphere Diameter	d_{ball}		
Ballast Sphere Volume	V_{ball}	Enclosed Chamber Mass	M_b+M_{ca}
Ballast Sphere Mass	M_{ball}	Chamber Content Target Mass	M_T
Number of Required Ballast Spheres	N_{balls}		
Float Ballast Cost	Cost	Water Temperature in Degrees °Kelvin	T_w
		Mean English Surface water Temperature in °C	T_{ang}
Density of Stainless Steel	ρ_{ss}	Dynamic Viscosity of Water at T_w °K	μ_w
Diameter of Wire	d_{wire}	Differential Mass	ΔMass
Radius of Wire	r_{wire}	Differential Density	$\Delta\rho$
Volume of Wire	V_{wire}	Actual Float Mass	M_{act}
Length of Wire	L_{wire}	Actual Float Mean Density	ρ_{act}
		Terminal Settling Velocity	v_t
		Required Terminal Settling Velocity (limit)	$vreq_t$

Table Two
Classified General Notation

Air density is at STP
Water Density is at 20 degrees C
Ping-pong balls are assumed to be made of cellulose nitrate
    and to be the International sporting standard 40mm diameter
    of 2.7 grams weight with a Coefficient of Restitution of 0.4
The density of epoxy putty is based upon a 2 US ounce cylindrical stick
    of "Quiksteel" as a 4-inch cylinder of 1.0625 inches diameter
    (http://www.dual-star.com/index2/Service/QuikSteel%20Epoxy%20Putty.htm)
The density of dry epoxy paint is computed from knowledge that
    ordinary Chinese epoxy ship paint is 1450kg/cu.m wet
    (Compartment Epoxy Paint (GLC-EPTANK 366))
    and that Akzo-Nobel Intershield 803Plus EGA-903Red
    Abrasion Resistant Binary Epoxy Paint is designed to leave a
    125 micron coat at 5sq.m/liter for 71% dry solids (computed value is 62.5%)
    (http://www.international-marine.com/PDS/4313+M+eng+A4.pdf)
Ballast Wire and Sphere Materials:-
    (a) Nickel Wire 99% 0.5mm dia Annealed NI005151
        5 meters = £80
        (http:/www.goodfellow.com/E/Nickel-Wire.html)
    (b) Nickel Sphere 99% 0.76mm dia NI006811
        50 spheres = £66.50
        (http:/www.goodfellow.com/E/Nickel-Sphere.html)
    (c) Nickel Wire 200 99% 0.39mm dia Annealed and Cold-drawn in Spool
        100 meters = £45
        (http://www.alloyshop.com/products/039mm-dia-nickel-200)
    (d) Nickel Wire 0.5mm dia 25 SWG Ref.: NI0500-050
        ~28 meters = 50grams = £11.45exVAT%
        (http://wires.co.uk/acatalog/ni_bare.html)
    (e) Stainless Steel 0.5mm dia Bare Ref.: SS0500-050
        50grams = £4exVAT%
        (http://wires.co.uk/acatalog/)

Table Three
Notes upon Materials and Components

Substance	Density ρ (kg/m^3)	Error Bound (+/- Kgs)
Air	1.2041	
Mercury	13546.0000	
Lead	11340.0000	
Tin	7365.0000	
Nickel	8908.0000	
Titanium	4506.0000	
Stainless Steel	7900.0000	
Glass	2579.0000	
Water	998.0000	
Cellulose Nitrate	1375.0000	25
Epoxy Putty	975.5889	
Dry Paint	1721.2000	

Table Four
Densities of Materials

A Mathematical and Physical Resolution of a
Cubic Equation for the Design of
Neutrally-Buoyant Liquid Ballasted Floats

by
James R Warren BSc MSc PhD PGCE

In my 10 November 2011 disquisition entitled "Aspects of the Design of Neutrally-Buoyant Liquid Ballasted Tachometric Floats" I explored the design load of liquid ballast needed in a ping-pong ball float, in the presence of residual chamber gas.

Whilst that work erected and solved a cubic equation for the requisite ballast depth it left the internal statics and algebra definitive of that desideratum unexplored.

This treatment develops that science and exposes both simplifications and limitations of the computation of design ballasted depth.

PART I
DATA

Certain essential data, the same as that for "Aspects", needs to be resolved.

The data is stated in Table One, where Input Data is noted in red and Output Data in Blue.

Mass of Displaced Stream Water

This determines the upthrust to be balanced and is given by:-

$$M_D = \rho_w \left(\frac{4}{3} \pi r^3 + V_l \right)$$

Equation 1

Chamber Content Target Mass

This is the increment of float mass required to balance the Displaced Mass:-

$$M_T = M_D - M_{SP}$$

Equation 2

Symbol	Description	Value
ρ_b	Ballast Density: Ballast is Liquid Mercury	13546.00000000
ρ_{ca}	Contained Gas Density: Contained Gas is Dry Air	1.20410000
ρ_w	Density of Stream Water	998.00000000
ρ_s	Density of Shell: Shell is Cellulose Nitrate	1375.00000000
$\Delta\rho_\alpha$	Chamber Content Differential Density	13544.79590000
ΔM_C	Whole-Chamber Mass Diffferential	0.42729341
r_s	External Shell Radius	0.02000000
r_p	Float Radius with Whole-Body Paint Layer	0.02012500
r_{ca}	Internal Radius of Float Chamber	0.01960146
t_s	Shell Thickness	0.00039854
V_l	Volume of Label Paint Layer	0.00000005
V_s	Volume of Shell	0.00000196
M_S	Mass of Shell	0.00270000
V_C	Float Chamber Volume	0.000031546685275
M_{SP}	Float Structure Total Body Mass	0.003873794668483
M_D	Mass of Displaced Stream Water	0.034124190190375
M_T	Chamber Content Target Mass	0.030250395521892
M_F	Mass of Air in Full Chamber with No Ballast	0.000037985363739

Table One
Extrinsic and Computed Solution Data

<u>Chamber Content Differential Density</u>

This is defined by:-

$$\Delta\rho_\alpha = \rho_b - \rho_{ca}$$
Equation 3

and is positive.

<u>Shell-Dependent Variables</u>

Firstly, to compute the Volume of the actual Shell Material, i.e. the cellulose nitrate hollow sphere, use:-

$$V_s = \frac{M_s}{\rho_s}$$
Equation 4

The Chamber Internal Radius is then given by:-

$$r_{ca} = \sqrt[3]{r_s^{\,3} - \frac{3V_s}{4\pi}}$$

Equation 5

Chamber Volume is thus:-

$$V_C = \frac{4}{3}\pi r_{ca}^{\,3}$$

Equation 6

From which the Whole-Chamber Mass Differential is given by:-

$$\Delta M_C = V_C\left(\rho_b - \rho_{ca}\right)$$
$$= V_C.\Delta\rho_\alpha$$
$$= \frac{4}{3}\pi r_{ca}^{\,3}.\Delta\rho_\alpha$$

Equation 7

<u>Mass of Air Contained in the Full Chamber with No Ballast</u>

This is given by:-

$$M_F = \frac{4}{3}\pi r_{ca}^{\,3}.\rho_{ca}$$

Equation 8

PART II
THE CUBIC EQUATION FOR CENTRAL DEPTH

If the design Central Depth of the Liquid Ballast Cap is noted as d, then the required Solution Cubic Equation is specified by:-

$$0 = k_0 + k_1 d + k_2 d^2 + k_3 d^3$$

Equation 9

where the Coefficients k_j are yielded by:-

$$K_0 = M_F - M_T \qquad \textbf{Equation 10a}$$
$$K_1 = 0 \qquad \textbf{Equation 10b}$$
$$K_2 = \pi r_{ca} \Delta \rho_\alpha \qquad \textbf{Equation 10c}$$
$$K_3 = \pi \frac{-\Delta \rho_\alpha}{3} \qquad \textbf{Equation 10d}$$

<u>The Discriminant Δ</u>

Should the Discriminant prove positive then this cubic equation shall yield three real roots.

The Discriminant is given by:-

$$\Delta = 18 k_0 k_1 k_2 k_3 - 4 k_0 k_2{}^3 + k_1{}^2 k_2{}^2 - 4 k_1{}^3 k_3 - 27 k_0{}^2 k_3{}^2$$

Equation 11

Because $k_1=0$ this reduces to:-

$$\Delta = -4 k_0 k_2{}^3 - 27 k_0{}^2 k_3{}^2$$

Equation 12

The substitution of mechanical terms then realises the Discriminant as:-

$$\Delta = -4 \left(\pi r_{ca} \Delta \rho_\alpha \right)^3 \left(M_F - M_T \right) - 27 \left(\pi \frac{-\Delta \rho_\alpha}{3} \right)^2 \left(M_F - M_T \right)^2$$

Equation 13

which simplifies to:-

$$\Delta = \pi^2 \Delta \rho_\alpha \left(M_T - M_F \right) \left[4 \pi r_{ca}{}^3 \Delta \rho_\alpha - 3 \left(M_T - M_F \right) \right]$$

Equation 14

For our data the value of D is in the region +65,167,402.82307635.
Therefore there exist three real roots.

The General System

Define the General System Factor, C_F, using:-

$$C_F = k_2^{\,2} - 3k_1 k_3 = k_2^{\,2} = \left(\pi r_{ca}\Delta\rho_\alpha\right)^2$$

Equation 15

If C_F is positive the General System yields three roots. Because all squares are positive, C_F is perforce positive and the General System yields three roots.
In our context, C_F is in the region of +695,699.0736145069.

PART III
THE DEFINITION OF ROOT COFACTOR C

All three roots, though themselves real, implicate a Complex Cofactor, C.

C is itself a Complex Cube Root of a Real Component, R, and a Complex Component Q.

Q has a zero real component.

The Complex Part of Q is q.

<u>The Resolution of Q</u>

Q is defined by:-

$$Q = \sqrt{\left(2k_2{}^3 - 9k_1k_2k_3 + 27k_0k_3{}^3\right)^2 - 4C_F{}^3}$$

Equation 16

or:-

$$Q = \sqrt{\left(2k_2{}^3 - 9k_1k_2k_3 + 27k_0k_3{}^3\right)^2 - 4k_2{}^6}$$

Equation 17

Because $k_1=0$ this reduces to:-

$$Q = \sqrt{\left(2k_2{}^3 + 27k_0k_3{}^3\right)^2 - 4k_2{}^6}$$

Equation 18

Q has a complex value exceeding 594 million.

By substitution Q may be expressed physically as:-

$$Q = \sqrt{\left[2\left(\pi r_{ca}\Delta\rho_\alpha\right)^3 + 27\left(\pi\frac{-\Delta\rho_\alpha}{3}\right)^2\left(M_F - M_T\right)\right]^2 - 4\left(\pi r_{ca}\Delta\rho_\alpha\right)^6}$$

Equation 19

This is progressively simplified as:-

$$Q = \sqrt{\pi^4 \Delta\rho_\alpha{}^4 \left[2\pi r_{ca}{}^3 \Delta\rho_\alpha + 3(M_F - M_T)\right]^2 - \pi^4 \Delta\rho_\alpha{}^4 \left(2\pi\Delta\rho_\alpha r_{ca}{}^3\right)^2}$$

$$= \pi^2 \Delta\rho_\alpha{}^2 \sqrt{9\left[\frac{4}{3}\pi r_{ca}{}^3 \Delta\rho_\alpha (M_F - M_T) + (M_F - M_T)^2\right]}$$

$$= 3\pi^2 \Delta\rho_\alpha{}^2 \sqrt{\left[\frac{4}{3}\pi r_{ca}{}^3 \Delta\rho_\alpha (M_F - M_T) + (M_F - M_T)^2\right]}$$

Equation 20

Because (M_F-M_T) is invariably negative and ΔM_C is much greater than (M_F-M_T), Q is always, as aforenoted, Complex of form 0+yi.

By substitution and insertion of the root of -1 we may summarise Q as:-

$$Q = 3\pi^2 \Delta\rho_a{}^2 \sqrt{\Delta M_C + M_F - M_T} \cdot \sqrt{M_T - M_F} \, .i$$

Equation 21

from which it is obvious that the Real Part of Q, q, is:-

$$q = 3\pi^2 \Delta\rho_a{}^2 \sqrt{\Delta M_C + M_F - M_T} \cdot \sqrt{M_T - M_F}$$

Equation 22

<u>The Resolution of R</u>

R is defined by:-

$$R = 2k_2{}^3 - 9k_1 k_2 k_3 + 27 k_0 k_3{}^2$$

Equation 23

By deletion of the term in k_1 this becomes:-

$$R = 2k_2{}^3 + 27 k_0 k_3{}^2$$

Equation 24

In context, the value of R is near to 996 million; just shy of one milliard.
By substitution, a mechanical expression of R is:-

$$R = 2\left(\pi r_{ca} \Delta\rho_\alpha\right)^3 + 27\left(\pi\frac{-\Delta\rho_\alpha}{3}\right)^2 (M_F - M_T)$$

Equation 25

Progressive simplification yields:-

$$R = \pi^2 \Delta\rho_\alpha{}^2 \left[2\pi r_{ca}{}^3 \Delta\rho_\alpha + 3\left(M_F - M_T\right) \right]$$

$$= \pi^2 \Delta\rho_\alpha{}^2 \left[\frac{3}{2}\Delta M_C + 3\left(M_F - M_T\right) \right]$$

$$= 3\pi^2 \Delta\rho_\alpha{}^2 \left[\frac{1}{2}\Delta M_C + \left(M_F - M_T\right) \right]$$

Equation 26

<u>The Composition of Root Cofactor C</u>

C is defined by:-

$$C = \sqrt[3]{\frac{1}{2}\left(Q + R\right)} = \sqrt[3]{\frac{1}{2}\left(R + q.i\right)} = \sqrt[3]{\frac{1}{2}\sqrt{R + q.i}}$$

Equation 27

After insertion of the physical expression of R we obtain:-

$$C = \sqrt[3]{\frac{1}{2}\left(Q + 2\left(\pi r_{ca}\Delta\rho_\alpha\right)^3 + 27\left(\pi \frac{-\Delta\rho_\alpha}{3}\right)^2 \left(M_F - M_T\right) \right)}$$

Equation 28

Then after the insertion of the physical expression of Q we obtain:-

$$C = \sqrt[3]{\frac{1}{2}\left\{ 3\pi^2 \Delta\rho_\alpha{}^2 \sqrt{\left[V_C \Delta\rho_\alpha \left(M_F - M_T\right) + \left(M_F - M_T\right)^2 \right]} + \left(\pi^2 \Delta\rho_\alpha{}^2\right)\left[2\pi r_{ca}{}^3 \Delta\rho_\alpha + 3\left(M_F - M_T\right) \right] \right\}}$$

Equation 29

Further simplification yields:-

$$C = \sqrt[3]{\frac{3\pi^2 \Delta\rho_\alpha{}^2}{2}\left\{ \left[\left(M_F - M_T\right) + \sqrt{\left(M_F - M_T\right)\left[\left(M_F - M_T\right) + \Delta M_C\right]}\right] + \frac{\Delta M_C}{2} \right\}}$$

Equation 30

<u>The Three Roots</u>

The three real roots for central ballast depth each comprise one real part, which proves to be r_{ca}, and two complex augends.

Tests of Absurdity

In general, only one of the three computed roots will be valid though all satisfy the cubic equation at a null point.

This is because two roots will either be negative or exceed the available chamber diameter. Accordingly:-

$$0 < d \le 2r_{ca} \qquad \text{for validity}$$

Indeed, for dense fluids like mercury the Central Depth d will be less than the Chamber Radius r_{ca} such that $d = x_2$.

Certain light liquids including water infill will exhaust the lower hemisphere and top toward the chamber zenith, leaving a gaseous cap. In such a circumstance, liquid Central Depth is interpreted as $d = 2r_{ca} - x_2$.

The First Root

The First Root, x_1, is given by:-

$$x_1 = \frac{\pi r_{ca} \Delta\rho_\alpha}{3\left(\pi \times \dfrac{-\Delta\rho_\alpha}{3}\right)} - \frac{C}{3\left(\pi \times \dfrac{-\Delta\rho_\alpha}{3}\right)} - \frac{C_F}{3\left(\pi \times \dfrac{-\Delta\rho_\alpha}{3}\right) \times C}$$

Equation 31

which reduces to:-

$$x_1 = r_{ca} + \frac{1}{\pi.\Delta\rho_\alpha}\left(C + \frac{C_F}{C}\right)$$

Equation 32

or:-

$$x_1 = r_{ca} + \frac{1}{\pi.\Delta\rho_\alpha}\left(C + \frac{\left(\pi r_{ca}\Delta\rho_\alpha\right)^2}{C}\right)$$

Equation 33

When C exceeds unity it is clear that the depth of Equation Thirty-Three must exceed $2 \times r_{ca}$, the Internal Diameter of the float: The First Root is accordingly absurd.

In our example, $d \equiv x_1$ is about 0.058175 meters.

The Second Root

The Second Root may be specified in mechanical terms as:-

$$x_2 = r_{ca} + \frac{C\left(1 + i\sqrt{3}\right)}{6\left(\pi \times \dfrac{-\Delta\rho_\alpha}{3}\right)} + \frac{C_F\left(1 - i\sqrt{3}\right)}{6\left(\pi \times \dfrac{-\Delta\rho_\alpha}{3}\right)}$$

Equation 34

which may be reduced to:-

$$x_2 = r_{ca} - \frac{1}{2\pi\Delta\rho_\alpha}\left[C\left(1 + i\sqrt{3}\right) + \frac{\left(1 - i\sqrt{3}\right)\left(\pi r_{ca}\Delta\rho_\alpha\right)^2}{C}\right]$$

Equation 35

For our data $x_2 \equiv d$ proves to be in the region of 0.006373822 meters or 6.373822 mm of cap Central Depth from which the instilled mass of mercury ballast is easily computed. x_2 is the only physically cogent estimate of Central Depth.

The Third Root

The Third Root may be specified in mechanical terms as:-

$$x_2 = r_{ca} + \frac{C\left(1 - i\sqrt{3}\right)}{6\left(\pi \times \dfrac{-\Delta\rho_\alpha}{3}\right)} + \frac{C_F\left(1 + i\sqrt{3}\right)}{6\left(\pi \times \dfrac{-\Delta\rho_\alpha}{3}\right)}$$

Equation 36

which may be reduced to:-

$$x_2 = r_{ca} - \frac{1}{2\pi\Delta\rho_\alpha}\left[C\left(1 - i\sqrt{3}\right) + \frac{\left(1 + i\sqrt{3}\right)\left(\pi r_{ca}\Delta\rho_\alpha\right)^2}{C}\right]$$

Equation 37

x_2 is negative and accordingly meaningless as a central depth.

PART IV
ANALYSIS OF THE ROOT COMPLEX ARGUMENT AND MODULUS

We saw in Part III that the Complex Cofactor C is defined by:-

$$C = \sqrt[3]{\frac{R+Q}{2}} = \sqrt[3]{\frac{R}{2} + \frac{Q}{2}}$$

Equation 38

Expressed in terms of the Real Components R and q, the Argument of C is given
by:-

$$\theta = \operatorname{atan}\left(\frac{\frac{q}{2}}{\frac{R}{2}}\right) = \operatorname{atan}\frac{q}{R}$$

Equation 39

and the Complex Argument, θ, is given by:-

$$\theta = \arg(R+Q)$$

Equation 40

Both the Real and Complex forms of θ are identical.
The Real Part Modulus, r, is given by:-

$$r = \sqrt{\left(\frac{R}{2}\right)^2 + \left(\frac{q}{2}\right)^2}$$

Equation 41

which is numerically identical to the Complex Modulus yielded by:-

$$r = \sqrt{\left(\frac{R}{2}\right)^2 + \left(\frac{Q}{2}\right)^2}$$

Equation 42

<u>Resolution of The Argument of C, θ</u>

Note that the mechanical forms of Q and R are given by:-

$$q = 3\pi^2 \Delta\rho_a{}^2 \sqrt{\Delta M_C + M_F - M_T} \cdot \sqrt{M_T - M_F}$$
Equation 22

and:-

$$R = \pi^2 \Delta\rho_\alpha{}^2 \left[2\pi r_{ca}{}^3 \Delta\rho_\alpha + 3(M_F - M_T) \right]$$
Equation 26

Using these forms we can by substitution and cancellation resolve a mechanical expression for θ as:-

$$\theta = \operatorname{atan}\frac{q}{R} = \operatorname{atan}\frac{3\pi^2 \Delta\rho_a{}^2 \sqrt{\Delta M_C + M_F - M_T} \cdot \sqrt{M_T - M_F}}{\pi^2 \Delta\rho_\alpha{}^2 \left[2\pi r_{ca}{}^3 \Delta\rho_\alpha + 3(M_F - M_T) \right]}$$
Equation 43

and thus:-

$$\theta = \operatorname{atan}\frac{3\sqrt{\Delta M_C + M_F - M_T} \cdot \sqrt{M_T - M_F}}{\left[2\pi r_{ca}{}^3 \Delta\rho_\alpha + 3(M_F - M_T) \right]}$$
Equation 44

<u>Analysis of S</u>

Let S be the Complex Cube of C, i.e.:-

$$C = \sqrt[3]{\frac{R+Q}{2}} = \sqrt[3]{S}$$
Equation 45

or:-

$$S = C^3$$
Equation 46

and:-

$$S = \frac{R+Q}{2}$$
Equation 47

then by substitution of Equations Twenty-Six and Twenty for R and Q respectively:-

$$S = \frac{\pi^2 \Delta\rho_\alpha{}^2 \left[2\pi r_{ca}{}^3 \Delta\rho_\alpha + 3(M_F - M_T) \right] + 3\pi^2 \Delta\rho_\alpha{}^2 \sqrt{\frac{4}{3}\pi r_{ca}{}^3 \Delta\rho_\alpha (M_F - M_T) + (M_F - M_T)^2} \,.i}{2}$$

Equation 48

or:-

$$S = \frac{1}{2}\pi^2 \Delta\rho_\beta{}^2 \left(2\pi r_{ca}{}^3 \Delta\rho_\alpha + 3M_F - 3M_T + \sqrt{3M_F - 3M_T + 4\pi r_{ca}{}^3 \Delta\rho}.\sqrt{M_F - M_T}.\sqrt{3} \right)$$

Equation 49

Resolution of the Real and Imaginary Components of S then gives:-

$$\Re(S) = \Re\left(\frac{R+Q}{2}\right) = \frac{1}{2}\pi^2 \Delta\rho_\alpha{}^2 \left[2\pi r_{ca}{}^3 \Delta\rho_\alpha + 3(M_F - M_T) \right]$$

Equation 50

and:-

$$\Im(S) = \Im\left(\frac{R+Q}{2}\right) = \sqrt{3(M_F - M_T) + 4\pi r_{ca}{}^3 .\Delta\rho_\alpha} \left[\frac{1}{2}\pi^2 \Delta\rho_\alpha{}^2 \left(\sqrt{3}.\sqrt{M_F - M_T} \right) \right]$$

Equation 51

Therefore:-

$$\theta_S = \operatorname{atan}\left(\frac{\Im(S)}{\Re(S)} \right)$$

Equation 52

$$\theta_S = -\operatorname{atan}\left(\frac{\sqrt{3(M_F - M_T) + 4\pi r_{ca}{}^3 .\Delta\rho_\alpha} \left[\frac{1}{2}\pi^2 \Delta\rho_\alpha{}^2 \left(\sqrt{3}.\sqrt{M_F - M_T} \right) \right].i}{\frac{1}{2}\pi^2 \Delta\rho_\alpha{}^2 \left[2\pi r_{ca}{}^3 \Delta\rho_\alpha + 3(M_F - M_T) \right]} \right)$$

Equation 53

$$\theta_S = -\mathrm{atan}\left(\frac{\sqrt{3(M_F - M_T)\left[3(M_F - M_T) + 4\pi\, r_{ca}{}^3.\Delta\rho_\alpha\right].i}}{\left[2\pi\, r_{ca}{}^3\Delta\rho_\alpha + 3(M_F - M_T)\right]}\right)$$

Equation 54

And with regard to the Complex Modulus of S, r_t, we have:-

$$r_t = \sqrt{\left(\Re(S)\right)^2 + \left(\Im(S)\right)^2}$$

Equation 55

In the mechanical form r_t is:-

$$r_t = \sqrt{\left(\frac{1}{2}\pi^2\Delta\rho_\alpha{}^2\left[2\pi\, r_{ca}{}^3\Delta\rho_\alpha + 3(M_F - M_T)\right]\right)^2 \ldots \atop \ldots + \left(\sqrt{3(M_F - M_T) + 4\pi\, r_{ca}{}^3.\Delta\rho_\alpha}\cdot\left[\frac{1}{2}\pi^2\Delta\rho_\alpha{}^2\left(\sqrt{3}.\sqrt{M_F - M_T}\right)\right]\right)^2}$$

Equation 56

This modulus can be expanded to:-

$$r_t = \sqrt{\pi^6 r_{ca}{}^6\Delta\rho_\alpha{}^6(M_F - M_T) + 3\pi^5 r_{ca}{}^3\Delta\rho_\alpha{}^5(M_F - M_T) + \frac{9}{4}\pi^4\Delta\rho_\alpha{}^4(M_F - M_T)^2 + \ldots \atop \ldots\left[\frac{9}{4}\pi^4\Delta\rho_\alpha{}^4(M_F - M_T)^2 + 3\pi^5 r_{ca}{}^3\Delta\rho_\alpha{}^5(M_F - M_T)\right]}$$

Equation 57

This object condenses to:-

$$r_t = \sqrt{\left[\pi^6 r_{ca}{}^6\Delta\rho_\alpha{}^6(M_F - M_T) + 3\pi^5 r_{ca}{}^3\Delta\rho_\alpha{}^5(M_F - M_T)\right] - \left[3\pi^5 r_{ca}{}^3\Delta\rho_\alpha{}^5(M_F - M_T)\right]}$$

Equation 58

which simplifies to:-

$$r_t = \sqrt{\pi^6 r_{ca}{}^6\Delta\rho_\alpha{}^6}$$

Equation 59

from which:-

$$r_t = \left(\pi\, r_{ca} \Delta \rho_\alpha \right)^3$$
Equation 60

<u>The Cube Root of the Complex Modulus of S</u>

Allow that the Cube Root of the Complex Modulus of S is c_{rt}, then:-

$$c_{rt} = \sqrt[3]{r_t}$$
Equation 61

and using Equation Sixty note that:-

$$c_{rt} = \pi\, r_{ca} \Delta \rho_a$$
Equation 62

Furthermore, regard these alternate forms of C_{rt} in terms of the real parts R and q:-

$$c_{rt} = \left(\frac{R^2 + q^2}{4} \right)^{\frac{1}{6}} \qquad\qquad \textbf{Equation 63a}$$

$$c_{rt} = \left(\frac{R^2}{2^2} + \frac{q^2}{2^2} \right)^{\frac{1}{6}} \qquad\qquad \textbf{Equation 63b}$$

$$c_{rt} = \left(\left(\frac{R}{2} \right)^2 + \left(\frac{q}{2} \right)^2 \right)^{\frac{1}{6}} \qquad\qquad \textbf{Equation 63c}$$

$$c_{rt} = \left(\sqrt{ \left(\frac{R}{2} \right)^2 + \left(\frac{q}{2} \right)^2 } \right)^{\frac{1}{3}} \qquad\qquad \textbf{Equation 63d}$$

PART VI
COMPLEX EXPONENTIAL ANALYSIS OF COFACTOR C

<u>Alternate Forms of the Root Cofactors</u>

The three alternate exponential forms of C are given by:-

$$C = \sqrt[3]{r}.\exp\left[i.\left(\frac{1}{3}\theta \right) \right] \qquad \textbf{Equation 64a}$$

$$C = \sqrt[3]{r}.\exp\left[i.\left(\frac{1}{3}\theta + \frac{2}{3}\theta \right) \right] \qquad \textbf{Equation 64b}$$

$$C = \sqrt[3]{r}.\exp\left[i.\left(\frac{1}{3}\theta - \frac{2}{3}\theta \right) \right] \qquad \textbf{Equation 64c}$$

Of these, the first form, Equation Sixty-Four a, is numerically appropriate to our analysis.

<u>The Cofactor C</u>

Substitution of Equations Sixty-Three d and Equation Thirty-Nine into Equation Sixty-Four a allows us to write:-

$$C = \left(\sqrt{\left(\frac{R}{2} \right)^2 + \left(\frac{q}{2} \right)^2} \right)^{\frac{1}{3}}.\exp\left[i.\left(\frac{1}{3}\operatorname{atan}\left(\frac{\frac{q}{2}}{\frac{R}{2}} \right) \right) \right]$$

Equation 65

which, upon substitution of Equation Sixty-Two for the RHS coefficient, and simplification of the argument yields:-

$$C = \pi \, r_{ca}\Delta\rho_\alpha.\exp\left[i\left(\frac{1}{3}\cdot\operatorname{atan}\frac{q}{R} \right) \right]$$

Equation 66

or:-

$$C = \pi \, r_{ca}\Delta\rho_\alpha.\exp\left[i\left(\frac{\theta}{3} \right) \right]$$

Equation 66a

<u>Analysis of the Complex Cofactor η and its Real Part η_r</u>

Let us define the Argument of θ, i.e. the generative real ratio of its opposite and adjacent. to be η such that:-

$$\eta = i \cdot \frac{q}{R}$$

Equation 67

η has no real component. However, η has a real form defined as:-

$$\eta_r = \frac{q}{R}$$

Equation 68

The relationship between η and η_r is given by:-

$$\theta = \operatorname{atan} \eta_r$$

Equation 69

Using Equation Sixty-Eight we may specify η_r is mechanical terms as:-

$$\eta_r = \frac{3\pi^2 \Delta \rho_\alpha{}^2 \sqrt{\Delta M_C + M_F - M_T} \cdot \sqrt{M_T - M_F}}{3\pi^2 \Delta \rho_\alpha{}^2 \left[\frac{1}{2} \Delta M_C + \left(M_F - M_T \right) \right]}$$

Equation 70

Cancellation gives:-

$$\eta_r = \frac{\sqrt{\Delta M_C + M_F - M_T} \cdot \sqrt{M_T - M_F}}{\left[\frac{2}{3} \pi r_{ca}{}^3 \Delta \rho_\alpha + \left(M_F - M_T \right) \right]}$$

Equation 71

which can be re-arranged as:-

$$\eta_r = \frac{\sqrt{\Delta M_C + M_F - M_T} \cdot \sqrt{M_T - M_F}}{\left[\frac{\Delta M_C}{2} + \left(M_F - M_T \right) \right]}$$

Equation 72

Accordingly the Complex Ratio η is given by:-

$$\eta_r = 2i.\frac{\sqrt{\Delta M_C + M_F - M_T}.\sqrt{M_T - M_F}}{\left[\Delta M_C + 2\left(M_F - M_T\right)\right]}$$

Equation 73

<u>The Complex Cofactor C in Terms of the Complex Ratio η and its Real Part η_r</u>

In terms of the Complex Ratio η, the Complex Cofactor C may be specified as:-

$$C = \pi\, r_{ca}\Delta\rho_\alpha.\exp\left(\frac{\text{atanh}\,\eta}{3}\right)$$

Equation 74

By substitution of the hyperbolic inverse tangent we may re-write this as:-

$$C = \pi\, r_{ca}\Delta\rho_\alpha.\exp\left[\frac{1}{3}\frac{1}{2}\log_n\left(\frac{1+\eta}{1-\eta}\right)\right]$$

Equation 75

or:-

$$C = \pi\, r_{ca}\Delta\rho_\alpha.\exp\left[\frac{1}{6}\log_n\left(\frac{1+\eta}{1-\eta}\right)\right]$$

Equation 76

or by eliminating the logarithm:-

$$C = \pi\, r_{ca}\Delta\rho_\alpha.\exp\left[\left(\frac{1+\eta}{1-\eta}\right)^{\frac{1}{6}}\right]$$

Equation 77

PART VII
THE SECOND ROOT AS A FUNCTION OF ARGUMENT RATIO η

In order to facilitate our further analyses let us define the auxiliaries:-

$$\alpha = \Re\left[\left(\frac{1+\eta}{1-\eta}\right)^{\frac{1}{6}}\right] = \cos\left(\frac{\arg(\lambda)}{6}\right) = \cos\left(\frac{\theta}{3}\right)$$

Equation 78

and:-

$$\beta = \Im\left[\left(\frac{1+\eta}{1-\eta}\right)^{\frac{1}{6}}\right] = \sin\left(\frac{\arg\lambda}{6}\right) = \sin\left(\frac{\theta}{3}\right)$$

Equation 79

Also:-

$$\lambda = \frac{1+\eta}{1-\eta}$$

Equation 80

such that:-

$$\theta = \frac{\arg\lambda}{2}$$

Equation 81

<u>Management of the Internal Arithmetic of The Second Solution Root x_2</u>

In mechanical terms, the Second Root x_2 is of course the valid ballast pool Central Depth d.

The simplification of x_2 necessitates the elimination of redundant complex factors internal to it.

In order to progress this need allow that:-

$$f_a = (\alpha + \beta i)\left(1 + i\sqrt{3}\right)$$

Equation 82

and:-

$$f_b = \frac{1 - i\sqrt{3}}{\alpha + \beta i}$$

Equation 83

Define F as the complex sum of f_a and f_b, i.e.:-

$$F = f_a + f_b = (\alpha + \beta i)(1 + i\sqrt{3}) + \frac{1 - i\sqrt{3}}{\alpha + \beta i}$$

Equation 84

The complex multiplication of the RHS of Equation Eighty-Two allows us to write:-

$$f_a = (\alpha - \sqrt{3}\beta) + (\beta + \sqrt{3}\alpha)i$$

Equation 85

And complex division of the RHS of Equation Eighty-Three allows us to write:-

$$f_b = \frac{\alpha - \sqrt{3}\beta}{\alpha^2 + \beta^2} + \frac{-\beta - \sqrt{3}\alpha}{\alpha^2 + \beta^2}i$$

Equation 86

The complex summation of f_a and f_b gives us:-

$$F = \left[(\alpha - \sqrt{3}\beta) + \frac{\alpha - \sqrt{3}\beta}{\alpha^2 + \beta^2} \right] + \left[(\beta + \sqrt{3}\alpha) + \frac{-\beta - \sqrt{3}\alpha}{\alpha^2 + \beta^2} \right] \cdot i$$

Equation 87

which may be re-arranged as:-

$$F = \left[(\alpha - \sqrt{3}\beta)\left[1 + \frac{1}{\alpha^2 + \beta^2} \right] \right] + \left[(\beta + \sqrt{3}\alpha)\left[1 - \frac{1}{\alpha^2 + \beta^2} \right] \right] \cdot i$$

Equation 88

By the Pythagorean Identity it is notable that:-

$$\alpha^2 + \beta^2 = 1$$

Equation 89

and thus the real multiplier:-

$$1 + \frac{1}{\alpha^2 + \beta^2} = 2$$

Equation 90a

and the imaginary multiplier:-

$$1 - \frac{1}{\alpha^2 + \beta^2} = 0$$

Equation 90b

Decomposition of Equation Eighty-Eight then shows that:-

$$F = \Re(F) + \Im(F).i = F_R + F_I.i$$

Equation 91

where:-

$$F_R = \Re(F) = \left(\alpha - \sqrt{3}\beta\right)\left[1 + \frac{1}{\alpha^2 + \beta^2}\right] = 2\left(\alpha - \sqrt{3}\beta\right)$$

Equation 92

$$F_I = \Im(F) = \left(\beta + \sqrt{3}\alpha\right)\left[1 - \frac{1}{\alpha^2 + \beta^2}\right] = 0$$

Equation 93

For our data, the value of F_R is near to 1.349658.

The nullity of difference described by Equation Ninety b guarantees that the resulting F is real and that the valid second root (i.e. Central Depth) can be specified in wholly real terms.

Substitution in Equation Thirty-Seven permits us to write:-

$$d \equiv x_2 = r_{ca} - \frac{r_{ca}}{2} \cdot F$$

Equation 94

or:-

$$x_2 = r_{ca}\left(1 - \frac{1}{2} \cdot F\right)$$

Equation 95

which leads to:-

$$x_2 = r_{ca}\left(1 - \alpha + \sqrt{3}\beta\right)$$

Equation 96

where both α and β are real.
Equation Ninety-Six expands to:-

$$x_2 = r_{ca}\left[1 - \Re\left[\left(\frac{1+\eta}{1-\eta}\right)^{\frac{1}{6}}\right] + \sqrt{3}.\Im\left[\left(\frac{1+\eta}{1-\eta}\right)^{\frac{1}{6}}\right]\right]$$

Equation 97

By substitution into De Moivre's Theorem this last equation may be written:-

$$d = x_2 = r_{ca}\left[1 - \cos\frac{\arg\lambda}{6} + \sqrt{3}.\sin\frac{\arg\lambda}{6}\right]$$

Equation 98

PART VIII
THE SECOND ROOT IN EXPONENTIAL FORM

An equivalent approach to that of the preceding part, which, however, utilises complex exponential representations of the Angular Argument θ leads more directly and elegantly to an analogous trigonometric expression of Central Depth.
Note that:-

$$C = \pi r_{ca}\Delta\rho_\alpha .\exp\left[\frac{1}{6}\log_n\left(\frac{1+\eta}{1-\eta}\right)\right]$$

$$= \pi r_{ca}\Delta\rho_\alpha .\exp\left[\frac{1}{6}\log_n(\lambda)\right]$$

Equation 76

and that equivalently:-

$$C = \pi\, r_{ca}\Delta\rho_\alpha . \exp\left[i\left(\frac{1}{3}\theta \right) \right]$$

$$= \pi\, r_{ca}\Delta\rho_\alpha . \exp\left[i\left(\frac{\theta}{3} \right) \right]$$

Equation 66

Then by substitution of Equation Sixty-Six into Equation Thirty-Seven:-

$$x_2 = r_{ca} - \frac{\pi\, r_{ca}\Delta\rho_\alpha}{2\pi\,\Delta\rho_\alpha}\left[\exp\left(i.\frac{\theta}{3} \right)\left(1 + i\sqrt{3} \right) + \frac{\left(1 - i\sqrt{3} \right)}{\exp\left(i.\dfrac{\theta}{3} \right)} \right]$$

Equation 99

which simplifies to:-

$$x_2 = r_{ca} - \frac{r_{ca}}{2}\left[\exp\left(i.\frac{\theta}{3} \right)\left(1 + i\sqrt{3} \right) + \frac{\left(1 - i\sqrt{3} \right)}{\exp\left(i.\dfrac{\theta}{3} \right)} \right]$$

Equation 100

This equation can be re-cast in trigonometric terms as:-

$$x_2 = r_{ca} - \frac{r_{ca}}{2}\left\{ \begin{array}{l} \left[\left(\cos\frac{\theta}{3} - \sqrt{3}.\sin\frac{\theta}{3} \right) + \left(1.\cos\frac{\theta}{3} - \sqrt{3}.\sin\frac{\theta}{3} \right) \right] \\ + \left[\left(\sin\frac{\theta}{3} + \sqrt{3}.\cos\frac{\theta}{3} \right) + \left(-\sqrt{3}.\cos\frac{\theta}{3} - 1.\sin\frac{\theta}{3} \right) \right] \end{array} \right\}$$

Equation 101

Equation One Hundred and One reduces to:-

$$x_2 = r_{ca} - r_{ca}\left(\cos\frac{\theta}{3} - \sqrt{3}.\sin\frac{\theta}{3} \right)$$

Equation 102

or:-

$$x_2 = r_{ca}\left(1 - \cos\frac{\theta}{3} + \sqrt{3}.\sin\frac{\theta}{3}\right)$$

Equation 103

Equations Ninety-Eight and One Hundred and Three are Equivalent because:-

$$\frac{\theta}{3} = \frac{\arg\lambda}{6}$$

Equation 104

By substitution for θ, Equation One Hundred and Three may be elaborated in physical form as:-

$$x_2 = r_{ca}\left(\begin{array}{l} 1 - \cos\left(\dfrac{1}{3}\cdot\operatorname{atan}\left[\dfrac{\sqrt{\Delta M_C + M_F - M_T}.\sqrt{M_T - M_F}}{\dfrac{2\pi\,r_{ca}{}^3\Delta\rho_\alpha}{3} + \left(M_F - M_T\right)}\right]\right) \\[4ex] + \sqrt{3}.\sin\left(\dfrac{1}{3}\cdot\operatorname{atan}\left[\dfrac{\sqrt{\Delta M_C + M_F - M_T}.\sqrt{M_T - M_F}}{\dfrac{2\pi\,r_{ca}{}^3\Delta\rho_\alpha}{3} + \left(M_F - M_T\right)}\right]\right) \end{array}\right)$$

Equation 105

or more succinctly:-

$$x_2 = r_{ca}\left(\begin{array}{l} 1 - \cos\left(\dfrac{1}{3}\cdot\operatorname{atan}\left[\dfrac{\sqrt{\Delta M_C + M_F - M_T}.\sqrt{M_T - M_F}}{\dfrac{\Delta M_C}{2} + M_F - M_T}\right]\right) \\[4ex] + \sqrt{3}.\sin\left(\dfrac{1}{3}\cdot\operatorname{atan}\left[\dfrac{\sqrt{\Delta M_C + M_F - M_T}.\sqrt{M_T - M_F}}{\dfrac{\Delta M_C}{2} + M_F - M_T}\right]\right) \end{array}\right)$$

Equation 106

For our data the Central Depth is near to 0.006373822 meters. Substitution of Equation Ten a allows the further economy of:-

$$x_2 = r_{ca}\left(1 - \cos\left(\frac{1}{3}\cdot\text{atan}\left[\frac{\sqrt{\Delta M_C + k_0}\cdot\sqrt{-k_0}}{\dfrac{\Delta M_C}{2} + k_0}\right]\right) + \sqrt{3}.\sin\left(\frac{1}{3}\cdot\text{atan}\left[\frac{\sqrt{\Delta M_C + k_0}\cdot\sqrt{-k_0}}{\dfrac{\Delta M_C}{2} + k_0}\right]\right)\right)$$

Equation 107

Equation One Hundred and Five may be stated in terms of η_r as:-

$$x_2 = r_{ca}\left[1 - \cos\left(\frac{\text{atan}(\eta_r)}{3}\right) + \sqrt{3}.\sin\left(\frac{\text{atan}(\eta_r)}{3}\right)\right]$$

Equation 108

In order to effect efficiencies it is convenient to define a Real Argument Function ψ as:-

$$\psi = \frac{\text{atan}(\eta_r)}{3} = \text{atan}\left[\frac{\sqrt{\Delta M_C + k_0}\cdot\sqrt{-k_0}}{\left[\dfrac{\Delta M_C}{2} + k_0\right]}\right] = \text{atan}\left[\frac{\sqrt{\Delta M_C + M_F - M_T}\cdot\sqrt{M_T - M_F}}{\left[\dfrac{\Delta M_C}{2} + \left(M_F - M_T\right)\right]}\right]$$

Equation 109

Accordingly, Equation One Hundred and Six may be rendered as:-

$$x_2 = r_{ca}\left(1 - \cos\psi + \sqrt{3}.\sin\psi\right)$$

Equation 110

<u>The Square of the Second Root</u>

The Square of the Second Root is of course:-

$$x_2^{\;2} = \left[r_{ca}\left(1 - \cos\psi + \sqrt{3}.\sin\psi\right)\right]^2$$

Equation 111

This object expands to:-

$$x_2^{\;2} = 4r_{ca}^{\;2} - 2r_{ca}^{\;2}\cos\psi + 2\sqrt{3}r_{ca}^{\;2}\sin\psi - 2r_{ca}^{\;2}\left(\cos\psi\right)^2 - 2\sqrt{3}r_{ca}^{\;2}\cos\psi\sin\psi$$

Equation 112

The square can then more succinctly be specified as:-

$$x_2{}^2 = 2r_{ca}{}^2 \left[2 - \cos\psi \left(1 + \cos\psi\right) + \sqrt{3} \sin\psi \left(1 - \cos\psi\right) \right]$$

Equation 113

PART IX
THE COMPUTATION OF
CHAMBER BALLAST AND RESIDUAL AIR MASSES

<u>Contained Ballast</u>

By the equation for the volume of a spherical cap the Volume of Contained Ballast may be given as:-

$$V_b = \frac{\pi\,x_2^{\,2}}{3}\left(3r_{ca} - x_2\right)$$

Equation 114

and accordingly the Ballast Mass is:-

$$M_b = \rho_b . V_b$$

Equation 115

By substitution of Equations One Hundred and Thirteen and One Hundred and Ten in Equation One Hundred and Fourteen we obtain:-

$$V_b = \frac{\pi}{3}\left\{2r_{ca}^{\,2}\left[2 - \cos\psi(1+\cos\psi) + \sqrt{3}\sin\psi(1-\cos\psi)\right]\cdot\left[3r_{ca} - r_{ca}\left(1-\cos\psi+\sqrt{3}\sin\psi\right)\right]\right\}$$

Equation 116

Simplification of Equation One Hundred and Sixteen enables:-

$$V_b = \frac{2}{3}\pi\,r_{ca}^{\,3}\left\{\left[2 - \cos\psi(1+\cos\psi) + \sqrt{3}\sin\psi(1-\cos\psi)\right]\cdot\left[2+\cos\psi-\sqrt{3}\sin\psi\right]\right\}$$

Equation 117

The internal multiplication of bracketed expressions, followed by the requisite cancellations, enables us to form this:-

$$V_b = \frac{2}{3}\pi\,r_{ca}^{\,3}\cdot\left(4 - 3\cos\psi^2 - \cos\psi^3 + 3\cos\psi.\sin\psi^2 - 3\sin\psi^2\right)$$

Equation 118

Grouping the squares of the Sine and Cosine then yields:-

$$V_b = \frac{2}{3}\pi\, r_{ca}{}^3 \cdot \left[4 - 3(\cos\psi^2 + \sin\psi^2) - \cos\psi^3 + 3\cos\psi.\sin\psi^2\right]$$

Equation 119

Application of The Pythagorean Identity then gives:-

$$V_b = \frac{2}{3}\pi\, r_{ca}{}^3 \cdot \left[1 - \cos\psi^3 + 3\cos\psi.\sin\psi^2\right]$$

Equation 120

Should we elect to have an equivalent form for Cosine only we may have:-

$$V_b = \frac{2}{3}\pi\, r_{ca}{}^3 \cdot \left[1 - 4\cos\psi^3 + 3\cos\psi\right]$$

Equation 121

Substitution in Equation One Hundred and Fifteen then gives:-

$$M_b = \rho_b\,\frac{2}{3}\pi\, r_{ca}{}^3 \cdot \left[1 - 4\cos\psi^3 + 3\cos\psi\right]$$

Equation 122

<u>Residual Contained Air</u>

The Volume of Contained Air is the difference of Ballast Volume subtracted from the Chamber Volume so that:-

$$V_{ca} = \frac{4}{3}\pi\, r_{ca}{}^3 - \frac{\pi x_2{}^2}{3}\left(3r_{ca} - x_2\right)$$

Equation 123

This is capable of straight-forward simplification, but such might be unprofitable in that it would introduce a cube of x_2.

The reverse substitution of Ballast Volume then restores:-

$$V_{ca} = \frac{4}{3}\pi\, r_{ca}{}^3 - V_b$$

Equation 124

The substitution for V_b then gives:-

$$V_{ca} = \frac{4}{3}\pi \, r_{ca}{}^{3} - \frac{2}{3}\pi \, r_{ca}{}^{3} \cdot \left[1 - \cos\psi^{3} + 3\cos\psi.\sin\psi^{2}\right]$$

Equation 125

This simplifies to:-

$$V_{ca} = \frac{2}{3}\pi \, r_{ca}{}^{3} \cdot \left[2 - (1 - \cos\psi^{3} + 3\cos\psi.\sin\psi^{2})\right]$$

Equation 126

which is equivalent to:-

$$V_{ca} = \frac{2}{3}\pi \, r_{ca}{}^{3} \cdot \left[1 + \cos\psi^{3} - 3\cos\psi.\sin\psi^{2}\right]$$

Equation 127

As for Ballast Volume, an alternative in Cosine only is available and that may be written:-

$$V_{ca} = \frac{2}{3}\pi \, r_{ca}{}^{3} \cdot \left[1 + 4\cos\psi^{3} - 3\cos\psi\right]$$

Equation 128

The Mass of Residual Confined Air is given by:-

$$M_{ca} = \rho_{ca}.V_{ca}$$

Equation 129

Substitution of Equation One Hundred and Twenty-Eight in One Hundred and Twenty-Nine then gives:-

$$M_{ca} = \rho_{ca} \cdot \frac{2}{3}\pi \, r_{ca}{}^{3}\left[1 + 4\cos\psi^{3} - 3\cos\psi\right]$$

Equation 130

<u>The Target Mass</u>

You possibly recall from the first page of this exploration that the Mass of Displaced Stream Water, M_D, is given by:-

$$M_D = \rho_w \left(\frac{4}{3} \pi r^3 + V_l \right)$$

Equation 1

and that the Chamber Content Target Mass is a simple difference of that and the structural mass of the float carcase, M_{SP}.

$$M_T = M_D - M_{SP}$$

Equation 2

At hydrostatic equipoise it is further the case that:-

$$M_T = M_b + M_{ca}$$

Equation 131

This sum can be expanded by inward substitution of Equations One Hundred and Twenty-Two and One Hundred and Thirty to show:-

$$M_T = \rho_b \frac{2}{3} \pi r_{ca}^3 \cdot \left[1 - \cos\psi^3 + 3\cos\psi.\sin\psi^2 \right] + \rho_{ca} \cdot \frac{2}{3} \pi r_{ca}^3 \left[1 + \cos\psi^3 - 3\cos\psi.\sin\psi^2 \right]$$

Equation 132

From which it is clear that:-

$$M_T = \frac{2}{3} \pi r_{ca}^3 \cdot \left[\rho_b \left(1 - 4\cos\psi^3 + 3\cos\psi \right) + \rho_{ca} \left(1 + 4\cos\psi^3 - 3\cos\psi \right) \right]$$

Equation 133

The Neutrally-Buoyant Hollow Float:
Shot and Wire as Ballast

by
James R Warren BSc MSc PhD PGCE

This disquisition extends the exploration of the design of neutrally-buoyant spherical tachometric floats ("ping-pong ball" floats). It treats of the use of dense spheres ("shot") or of malleable wire as alternative ballasts within the float chamber. There is no liquid ballast.

We shall assume that the Total Carcase Mass, M_{SP}, of the float structure, a composite of the shell, paint and plug mass contributions is given as 0.003873794668483 kg, and that the Volume of the Label Paint Layer is 0.00000005 m^3.

Data

The definition and assumed values of other data are tabulated below:-

Symbol	Definition	Value
M_s	Mass of Shell	0.0027
M_{SP}	Float Structure Total Body Mass	0.0038738
ρ_w	Density of Stream Water	998
ρ_s	Density of Cellulose Nitrate	1375
ρ_{ca}	Density of Contained Gas (Dry Air)	1.2041
ρ_{ss}	Density of Stainless Steel	7900
ρ_{Pb}	Density of Lead	11340
ρ_{Sn}	Density of White Tin	7365
ρ_{Hg}	Density of Mercury	13546
Pb%	The Fraction of Lead in Pewter	37
Sn%	The Fraction of White Tin in Pewter	63
r_s	Radius of Float Shell	0.02
r_{ss}	Radius of Stainless Steel Wire	0.00025
r_p	Radius of Float with Whole-Body Paint Layer	0.020125
r_{Pb}	Radius of Spherical Shot (lead)	0.0005
r_{ball}	Radius of Spherical Shot (pewter)	0.000225
V_l	Volume of Label Paint Layer	5E-08

Table One
The Input Data Values and Definitions

Symbol	Definition	Value
L	Required Length of Wire	9.7401264
ΔM_C	Whole-Chamber Mass Differentilal	0.2491808
M_D	Mass of Displaced Stream Water	0.0341242
M_F	Mass of Air Contained in the Full Chamber with No Ballast	3.799E-05
M_T	Chamber Content Target Mass	0.0302504
M_{Pb}	Mass of Lead Shot Ballast	0.0302156
M_{SnPb}	Mass of Pewter Shot Ballast	0.0302167
M_{ss}	Mass of Wire Ballast	0.030217
n_{Pb}	Required Number of Lead Shot (fractional)	5088.8519
n_{SnPb}	Required Number of Pewter Shot (fractional)	74835.791
$\Delta\rho_{Pb}$	Chamber Content Differential Density (Lead)	11338.796
$\Delta\rho_{SnPb}$	Chamber Content Differential Density (Pewter)	8461.3538
$\Delta\rho_{ss}$	Chamber Content Differential Density (Stainless Steel)	7898.7959
ρ_{SnPb}	Density of Pewter	8462.5579
r_{ca}	Internal Radius of Float Chamber	0.0196015
t_s	Shell Thickness	0.0003985
V_{ca}	Internal Volume of Float Chamber	3.155E-05
V_C	Internal Volume of Float Chamber	3.155E-05
V_s	Volume of Shell Material	1.964E-06
V_{Pbball}	Volume of Spherical Lead Shot Ball	5.236E-10
$V_{SnPbball}$	Volume of Spherical Pewter Shot Ball	4.771E-11
V_{Pb}	Volume of All of the Lead Ballast Shot	2.665E-06
V_{SnPb}	Volume of All of the Pewter Ballast Shot	3.571E-06
V_{pewter}	Volume of Pewter (normative for density calculation)	0.0118168
V_{ss}	Volume of Required Stainless Steel	3.825E-06
Vpt_{Pb}	Fractional Volume of Lead	0.0032628
Vpt_{Sn}	Fractional Volume of White Tin	0.008554
W%	The Sum of Alloy Component Fractions	100
ρ_b	Generalised Density of Ballast	
$\Delta\rho_{Pb}$	Generalised Chamber Content Differential Density	

Table Two
The Output Data Values and Definitions

These figures are illustrative of the shot and wire ballasting arrangements. Note that the dimensions of the balls and wire are exaggerated, and their number and length understated. This is done to clarify the concept of ballasting with such media.

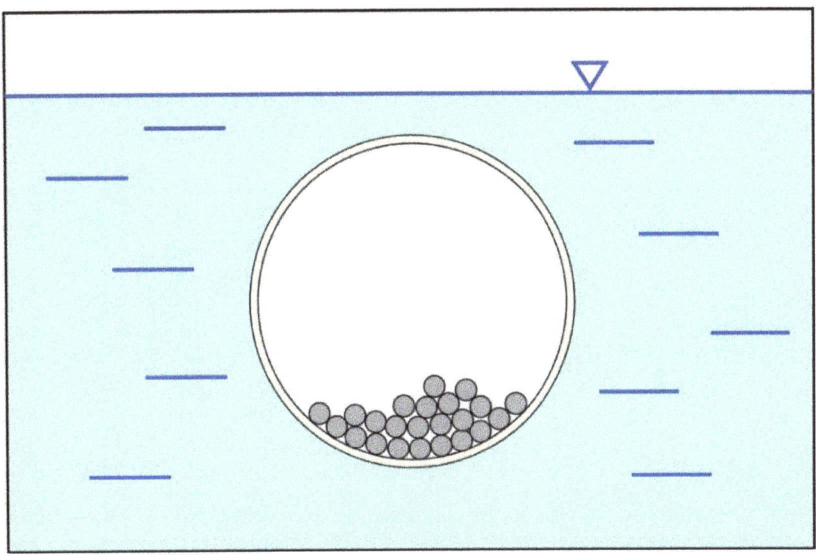

Figure One
The Shot Ballasting Arrangement
in Principle

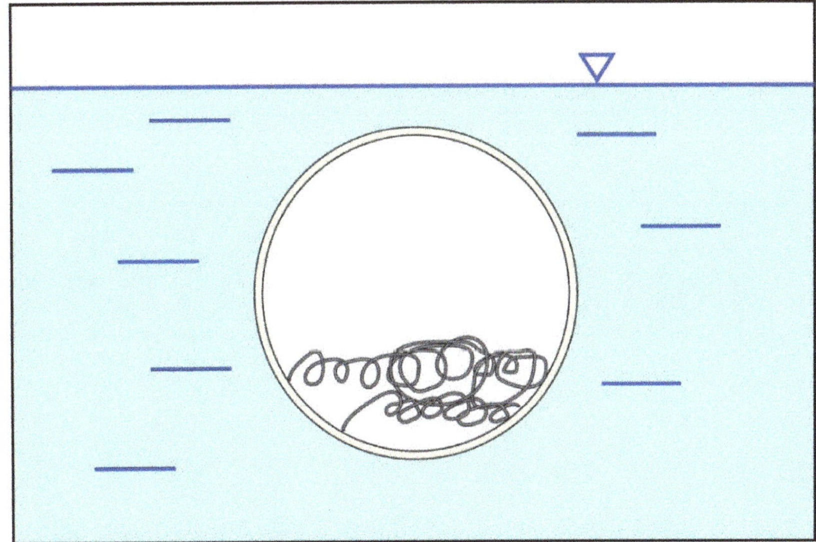

Figure Two
The Wire Ballasting Arrangement
in Principle

The Mass of the Displaced Stream Water, M_D, is given by:-

$$M_D = \rho_w \left(\frac{4}{3} \pi r_p^{\,3} + V_l \right)$$

Equation 1

and Chamber Target Mass, M_T, by:-

$$M_T = M_D - M_{SP}$$

Equation 2

The Volume of the Shell Material, V_s, is yielded by:-

$$V_s = \frac{M_s}{\rho_s}$$

Equation 3

from which it follows that the Inner Radius of the Chamber is given by:-

$$r_{ca} = \sqrt[3]{r_s^{\,3} - \frac{3V_s}{4\pi}}$$

Equation 4

Shell Thickness is computed from this using:-

$$t_s = r_s - r_{ca}$$

Equation 5

The Float Chamber Volume, V_{ca}, (sometimes noted as V_C) is also a function of Chamber Radius, r_{ca}.
It is yielded by:-

$$V_{ca} = \frac{4}{3} \pi r_{ca}^{\,3}$$

Equation 6

The Volume of the Shell Material (i.e. the cellulose nitrate envelope plus the epoxy plug and any whole-surface external paint) is given by:-

$$V_s = \frac{4}{3}\pi r_s^{\,3} - \frac{4}{3}\pi r_{ca}^{\,3} = \frac{4}{3}\pi\left(r_s^{\,3} - r_{ca}^{\,3}\right)$$

Equation 7

The generalised Chamber Content Density Differential is:-

$$\Delta\rho = \rho_b - \rho_{ca}$$

Equation 8

using which we may compute the Whole-Chamber Mass Differential, ΔM_C, as:-

$$\Delta M_C = V_{ca}.\Delta\rho = \frac{4}{3}\pi r_{ca}^{\,3}.\Delta\rho$$

Equation 9

The Mass of Air in the Full Chamber with No Ballast is:-

$$M_F = \frac{4}{3}\pi r_{ca}^{\,3}.\rho_{ca}$$

Equation 10

<u>The Requisite Number of Lead Shot</u>

For trial purposes we make an arbitrary assumption that the shot is pure lead with a ball radius of 0.5 mm. Commercial nickel or solder balls are other possibilities and at a practical level the tendency of lead to patinate might cause uncertainty about the actual density of its comminuted forms.

The Mass of the necessary Lead Shot Ballast, M_{Pb}, is given by:-

$$M_{Pb} = \rho_{Pb}.n.\frac{4}{3}\pi r_{Pb}^{\,3}$$

Equation 11

whilst the Mass of the Residual Chamber Air, M_{ca}, is:-

$$M_{ca} = \rho_{ca}\left(V_{ca} - V_{Pb}\right) = \rho_{ca}.\frac{4}{3}\pi\left(r_{ca}^{\,3} - n.r_{Pb}^{\,3}\right)$$

Equation 12

The Target Mass for hydrostatic balance is the sum of these two contained masses:-

$$M_T = M_{Pb} + M_{ca} = \rho_{Pb}.V_{Pb} + \rho_{ca}.\left(V_{ca} - V_{Pb}\right)$$

Equation 13

which may be expanded as:-

$$M_T = \frac{4}{3}\pi\left[n.\rho_{Pb}.r_{Pb}^{\ 3} + \rho_{ca}.r_{ca}^{\ 3} - \rho_{ca}.n.r_{Pb}^{\ 3}\right]$$

Equation 14

or:-

$$M_T = \frac{4}{3}\pi\left[n.r_{Pb}^{\ 3}\left(\rho_{Pb} - \rho_{ca}\right) + \rho_{ca}.r_{ca}^{\ 3}\right]$$

Equation 15

Re-arrangement of Equation Fifteen enables us to calculate the requisite Number of Shot Balls as:-

$$n_{Pb} = \frac{3M_T}{4\pi\,r_{Pb}^{\ 3}.\Delta\rho} - \frac{\rho_{ca}}{\rho_{Pb} - \rho_{ca}}\cdot\frac{r_{ca}^{\ 3}}{r_{Pb}^{\ 3}} = \frac{1}{\Delta\rho}\left(\frac{3M_T}{4\pi\,r_{Pb}^{\ 3}} - \rho_{ca}\frac{r_{ca}^{\ 3}}{r_{Pb}^{\ 3}}\right)$$

Equation 16

or:-

$$n_{Pb} = \frac{1}{\Delta\rho.r_{Pb}^{\ 3}}\left(\frac{3M_T}{4\pi} - \rho_{ca}r_{ca}^{\ 3}\right)$$

Equation 17

In this example n_{Pb} = 5088.851897640857.

It is obvious and unwelcome that the ballast discretisation has rendered precise hydrostatic equilibrium unattainable.

The Use of Scotle Pewter Balls

The Scotle firm manufactures tiny pewter balls intended for use as BGA solder balls for electrical fitting.

In November 2011 Internet supply of these products was available on Ebay. The 0.45mm diameter pewter balls were packed in 25000-round bottles retailing at about £2.12 post paid. The advertisement stated that the spherules were 37% Pb and 63% Sn, presumably by weight. I made the further assumption that the tin is of the white allotropy and mixes with the lead in simple mechanical promiscuity, uncomplicated by the segregation of any chemical product.

On such assumptions the Composite Density of Scotle Pewter may be written:-

$$\rho_{SnPb} = \frac{W\%}{V_{pewter}} = \frac{W\%}{Vpt_{Pb} + Vpt_{Sn}} = \frac{Pb\% + Sn\%}{\dfrac{Pb\%}{\rho_{Pb}} + \dfrac{Sn\%}{\rho_{Sn}}}$$

Equation 18

Approximate data yields a Pewter Density, ρ_{SnPb}, near to 8462.5579 kg/m^3. A useful version of Equation Sixteen may be specified as:-

$$n_{SnPb} = \frac{M_T}{\frac{4}{3}\pi r_{SnPb}{}^3\left(\rho_{SnPb} - \rho_{ca}\right)} - \frac{\rho_{ca}}{\rho_{SnPb} - \rho_{ca}} \cdot \frac{r_{ca}{}^3}{r_{SnPb}{}^3}$$

Equation 19

This computes as 74835.79 balls per float at a cost of about £6.35.

<u>The Requisite Length of Stainless Steel Wire</u>

For trial purposes we make the assumption that the wire is of stainless steel of radius 0.00025 meters (i.e. half-millimeter diameter). The density of this material is 7900 kg/m^3, at least for our demonstrations. The wire is clean, bright and untarnished. It is perfectly cylindrical and of constant gage.

The Mass of the necessary Wire Ballast, M_{ss}, is given by:-

$$M_{ss} = \rho_{ss}.2\pi r_{ss}{}^2.L$$

Equation 20

whilst the Mass of the Residual Chamber Air, M_{ca}, is:-

$$M_{ca} = \rho_{ca}\left(V_{ca} - V_{ss}\right) = \rho_{ca}\left(\frac{4}{3}\pi r_{ca}{}^3 - 2\pi r_{ss}{}^2 L\right)$$

Equation 21

It is of course possible to assemble the above equation as:-

$$M_{ca} = 2\pi\rho_{ca}\left(\frac{2}{3}r_{ca}{}^3 - r_{ss}{}^2 L\right)$$

Equation 22

The Target Mass for hydrostatic balance is the sum of these two contained masses:-

$$M_T = M_{ss} + M_{ca}$$

Equation 23

which may be expressed as:-

$$M_T = \rho_{ss} V_{ss} + \rho_{ca}\left(V_{ca} - V_{ss}\right)$$

Equation 24

which we will expand to be:-

$$M_T = \rho_{ss}.2\pi\, r_{ss}^{\,2}.L + \rho_{ca}.\frac{4}{3}\pi\, r_{ca}^{\,3} - \rho_{ca}.2\pi\, r_{ss}^{\,2} L$$

Equation 25

Re-arrangement of Equation Twenty-Five permits us to draft forms closely analogous to the shot count Equation Sixteen:-

$$L = \frac{M_T}{2\pi\, r_{ss}^{\,2}.\Delta\rho} - \frac{2}{3}.\frac{\rho_{ca}}{\Delta\rho}.\frac{r_{ca}^{\,3}}{r_{ss}^{\,3}}$$

Equation 26

where in context:-

$$\Delta\rho = \rho_{ss} - \rho_{ca}$$

Equation 27

A convenient arrangement for L is:-

$$L = \frac{M_T - \rho_{ca}\left(\dfrac{4}{3}\pi\, r_{ca}^{\,3}\right)}{2\pi\, r_{ss}^{\,2}.\Delta\rho}$$

Equation 28

For the given data the Requisite Wire Length L computes to be about 9.74014623 meters.

The Neutrally-Buoyant Hollow Float:
Shell Thickness Control to Ballast

by
James R Warren BSc MSc PhD PGCE

A hollow sphere may have its wall thickness adjusted in such a manner as to establish its neutral buoyancy in a fluid of given density, provided that the material of which it is made is denser than the suspensive medium.

The chamber within such a float contains no tertiary ballast component.

We may describe such a float as "monocoque" in the slightly specialised sense that the structure is a self-subsistent, and self-effective, unity that contains no inserted third ballast component, nor any plug. We do not, however, rule out external coatings. In this treatment we shall assume that the float is presented naked in order to simplify the algebraic and computational demonstrations.

A float of this sort may possibly be made using modern 3D printing (i.e. additive manufacturing). For the sake of consistency with our ping-pong ball float studies we shall assume our monocoque float wall material to be cellulose nitrate, with a Density ρ_s of 1375 kg/m^3, in the full knowledge that this material may be radically unsuitable.

More practical shell materials may include Acrylonitrile Butadiene Styrene (ABS) which has a density near to 1040 kg/m^3, or the biodegradable plastic Polycaprolactone with a density of about 1145 kg/m^3. Both lend themselves to additive manufacture, and in particular Fusion Deposition Modelling (FDM).

Merely to indicate some of what is out there these days I found that www.3Dprintingsystems.com is offering an entry-level UP! additive printer for US$2450 (January 2012) and this firm alleges that the user can build ABS artefacts for as little as US$0.08 per gram. Doubtless as prices decline such systems shall commend themselves to the small workshop, and built floats will be thinkable to the amateur and microhydro interests.

This development will neglect the contribution of float coatings and plug modifiers: There is no plug. Accordingly the Float Carcase Mass is not directly implicated, being implicit in the computed Float Mass, M_{float}, which is equivalent to the Mass of Displaced Water, M_D.

Figure One is schematic illustration of this arrangement, as a section through the (x,z) plane.

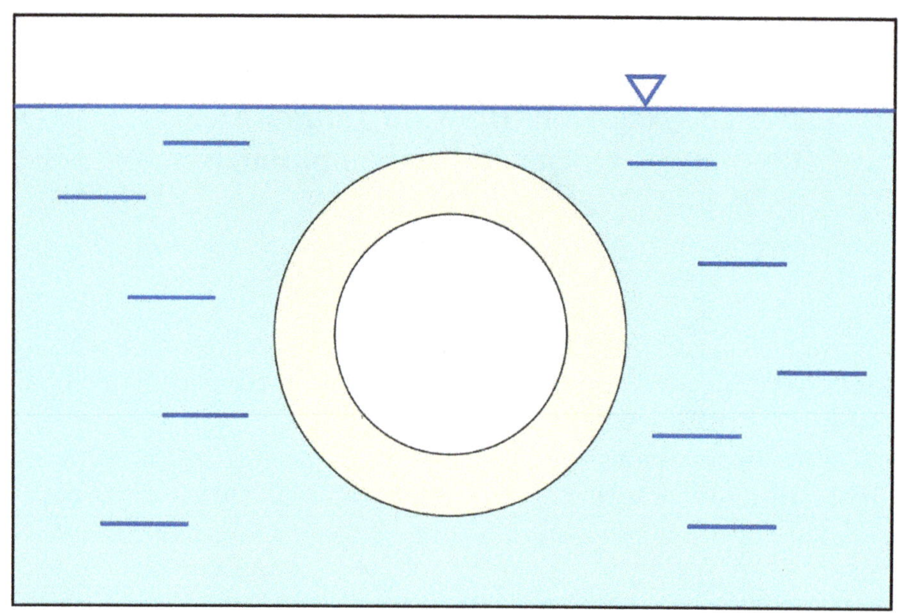

Figure One
A Design Section of the
Monocoque Neutrally-Buoyant Float

PART I
DATA

Table One lists the transput data notation and definitions, together with illustrative trial values. As is conventional, the input data values are in red and the output in blue.

Symbol	Definition	Value
t	Wall Thickness	0.007003202
r	Float Radius: r_s	0.02
s	Internal Float Radius: r_{ca}	0.012996798
ρ_w	Density of Stream Water	998
ρ_s	Density of Cellulose Nitrate	1375
ρ_{ca}	Density of Contained Gas (Dry Air)	1.2041
$\Delta\rho_\alpha$	Differential Density between Shell Material and Dry Air (positive)	1373.7959
$\Delta\rho_\beta$	Differential Density between Stream Water and Dry Air (negative)	-996.7959
$\Delta\rho_\gamma$	Differential Density between Stream Water and Dry Air (positive)	996.7959
M_T	Chamber Content Target Mass	0.033443301
M_D	Mass of Displaced Stream Water	0.033443301
M_{float}	Float Mass	0.033443301
x_1	The First Root of the Cubic Equation	0.007003202
x_2	The Second Root of the Cubic Equation	0.032997+0.022511i
x_3	The Third Root of the Cubic Equation	0.032997-0.022511i
V_s	Volume of Shell	2.43143E-05
V_{ca}	Volume of Contained Air	9.19597E-06
i	The Square Root of Minus One	i
k_0	Coefficient Zero of the Cubic Equation	-0.033402951
k_1	Coefficient One of the Cubic Equation	6.905451371
k_2	Coefficient Two of the Cubic Equation	-345.2725686
k_3	Coefficient Three of the Cubic Equation	5754.542809
Δ	Discriminant	-142700.4385
C_F	Squared Areal Mass Density Cofactor	0
C	Cube Root Cofactor	224.3718939
Q	Third Real Cofactor	11295497.43
R	Second Real Cofactor	11295497.43
P	First Real Cofactor	0

Table One
Input and Output Data and its Definitions

PART II
THE FORMATION OF THE SOLUTION CUBIC EQUATION

We require to resolve a Wall Thickness, t, which defines a spherical float whose Total Mass is the Target Mass, M_T, which is equivalent to the Displaced (Water) Mass, M_D.

We know the Float Radius, r, and may note that the Internal Chamber Radius, s, is given by:-

$$s = r - t$$

Equation 1

whilst:-

$$M_T = M_D = \rho_w \cdot \frac{4}{3} \pi r^3$$

Equation 2

Clearly, M_T may be expressed as:-

$$M_T = \rho_w \cdot \frac{4}{3} \pi (s + t)^3$$

Equation 3

as well as:-

$$M_T = \rho_s V_s + \rho_{ca} V_{ca}$$

Equation 4

Now the Volume of the Contained Chamber is the same as the Volume of Contained Air, V_{ca}, and is given by:-

$$V_{ca} = \frac{4}{3} \pi s^3$$

Equation 5

whilst the Volume of the Float Shell is the Volume of Cellulose Nitrate, V_s:-

$$V_s = \frac{4}{3} \pi r^3 - \frac{4}{3} \pi s^3 = \frac{4}{3} \pi \left(r^3 - s^3 \right)$$

Equation 6

Therefore by incorporation of Equations Five and Six into Equation Four the latter may be re-specified as:-

$$M_T = \rho_s \left(\frac{4}{3}\pi r^3 - \frac{4}{3}\pi s^3 \right) + \rho_{ca}.\frac{4}{3}\pi s^3$$

Equation 7

Simplification of Equation Seven will produce:-

$$M_T = \frac{4}{3}\pi \left[\rho_s r^3 - s^3 \left(\rho_s - \rho_{ca} \right) \right]$$

Equation 8

or:-

$$M_T = \frac{4}{3}\pi \left(\rho_s r^3 - \Delta\rho_\alpha s^3 \right)$$

Equation 9

At this stage it is useful to eliminate s by noting that:-

$$s = r^3 - 3r^2 t + 3rt^2 - t^3$$

Equation 10

so that by substitution:-

$$M_T = \frac{4}{3}\pi \left[\rho_s r^3 - \left(r^3 - 3r^2 t + 3rt^2 - t^3 \right).\Delta\rho_\alpha \right]$$

Equation 11

Expansion of Equation Eleven and simplification permit us to arrive at the cubic equation:-

$$M_T - \frac{4}{3}\pi r^3.\rho_{ca} + 4\pi r^2.t.\Delta\rho_\alpha - 4\pi r.t^2.\Delta\rho_u + \frac{4}{3}\pi t^3.\Delta\rho_\alpha$$

Equation 12

Because:-

$$0 = k_0 + k_1 t + k_2 t^2 + k_3 t^3$$

Equation 13

we are able inspectionally to define the four coefficients $k_0,...,k_3$ and organise solution.

PART III
THE SOLUTION FACTORS IN MECHANICAL TERMS

For purposes of algebraic summarisation it is convenient to note these Density Differentials:-

$$\Delta\rho_\alpha = \rho_s - \rho_{ca} \qquad \textbf{Equation 14a}$$

$$\Delta\rho_\beta = \rho_{ca} - \rho_w \qquad \textbf{Equation 14b}$$

$$\Delta\rho_\gamma = \rho_w - \rho_{ca} \qquad \textbf{Equation 14c}$$

$\Delta\rho_\alpha$ and $\Delta\rho_\gamma$ are positive whilst $\Delta\rho_\beta$ is negative for all hydrostatically-useful substances.

By study of the cubic equation, Equation Twelve, we can write the set of Cubic Coefficients as:-

$$k_0 = \pi.\rho_{ca}.\frac{4}{3}r^3 - M_T \qquad \textbf{Equation 15a}$$

$$k_1 = +4\pi r^2.\Delta\rho_\alpha \qquad \textbf{Equation 15b}$$

$$k_2 = -4\pi r.\Delta\rho_\alpha \qquad \textbf{Equation 15c}$$

$$k_3 = \frac{4}{3}\pi.\Delta\rho_\alpha \qquad \textbf{Equation 15d}$$

The Discriminant Δ

The Discriminant Δ is defined by:-

$$\Delta = 18k_0 k_1 k_2 k_3 - 4k_0 k_2^{\,3} + k_1^{\,2}k_2^{\,2} - 4k_3 k_1^{\,3} - 27k_0^{\,2}k_3^{\,2}$$

Equation 16

For our data, the value of Δ lies near to -142700.4384858641.

Since the discriminant is negative there is one real root, x_1, and two complex conjugate roots, x_2 and x_3.

The complex roots are of course mechanically absurd. Notwithstanding that, their sums and products are both real and might furnish useful intelligence.

Root Solution Cofactors

In order to facilitate root interpretations it is convenient to define a set of composite cofactors that may be combined to establish various intermediate terms.

In particular we wish to resolve the real Squared Areal Mass Density Cofactor, C_F; and the four real cofactors P, Q, R and C.

In this system C_F and P are the same defined as:-

$$C_F = P = k_2{}^2 - 3k_1 k_3$$

Equation 17

The substitution of the mechanical values of coefficients permits us to write:-

$$P = 16\pi^2 r^2 \Delta\rho_\alpha{}^2 - 3 \times \frac{4}{3}\pi.\Delta\rho_\alpha \times 4\pi r^2.\Delta\rho_\alpha$$
$$= 16\pi^2 r^2 \Delta\rho_\alpha{}^2 - 16\pi^2 r^2 \Delta\rho_\alpha{}^2$$
$$= 0$$

Equation 18

Therefore C_F and P are inherently absent (MATHCAD® returns P as 5.8208×10^{-11} due to computational error).

Meanwhile, R is defined by:-

$$R = 2k_2{}^3 - 9k_1 k_2 k_3 + 27k_0 k_3{}^2$$

Equation 19

whose value is about +11,295,497.43024179.

Q is defined by:-

$$Q = \sqrt{R^2 - 4P^3}$$

Equation 20

Because P is zero as above noted we have:-

$$Q = \sqrt{R^2} \equiv R$$

Equation 21

Lastly, C is given by:-

$$C = \sqrt[3]{\frac{Q+R}{2}}$$

Equation 22

Therefore, in this hollow monocoque float context:-

$$C = \sqrt[3]{R}$$
Equation 23

The Mechanical Expression of R

Because C is critical to roots definition it would pay to express the cofactor R in its simplest mechanical form.

By the substitution of all the Coefficients $k_0,...,k_3$ in Equation Nineteen we may present R in its full mechanical form as:-

$$R = 2\left(-4\pi r.\Delta\rho_\alpha\right)^3 - 9\left(\frac{4}{3}\pi.\Delta\rho_\alpha\right)\left(-4\pi r.\Delta\rho_\alpha\right)\left(4\pi r^2.\Delta\rho_\alpha\right) + 27\left(\frac{4}{3}\pi.\Delta\rho_\alpha\right)^2\left(\frac{4}{3}\pi r^3\rho_{ca} - M_T\right)$$
Equation 24

We shall defer the resolution of M_T to a convenient later stage.
Simplification allows us to reduce R to:-

$$R = 64\pi^3 r^3 \Delta\rho_\alpha{}^3 + 64\pi^3 r^3 \Delta\rho_\alpha{}^2.\rho_{ca} - 48\pi^2\Delta\rho_\alpha{}^2.M_T$$
Equation 25

and:-

$$R = \left(4\pi r\right)^3\left(\Delta\rho_\alpha{}^3 + \Delta\rho_\alpha{}^2.\rho_{ca}\right) + 3\left(4\pi.\Delta\rho_\alpha\right)^2.M_T$$
Equation 26

By the substitution of the Displaced Mass, M_T, preceded by the division of the first group terms by $\Delta\rho_\alpha{}^3$, we may re-write R as:-

$$R = \left(4\pi r.\Delta\rho_\alpha\right)^3\left(1 + \frac{\rho_{ca}}{\Delta\rho_\alpha}\right) - 3\left(4\pi.\Delta\rho_\alpha\right)^2\left(\rho_w\frac{4}{3}\pi r^3\right)$$
Equation 27

Equation Twenty-Seven further reduces to:-

$$R = \left(4\pi r.\Delta\rho_\alpha\right)^3\left(1 + \frac{\Delta\rho_\alpha{}^2\rho_{ca}}{\Delta\rho_\alpha{}^3} - \frac{\Delta\rho_\alpha{}^2\rho_w}{\Delta\rho_\alpha{}^3}\right)$$
Equation 28

or:-

$$R = \left(4\pi r.\Delta\rho_\alpha\right)^3\left(1 + \frac{\rho_{ca}}{\Delta\rho_\alpha} - \frac{\rho_w}{\Delta\rho_\alpha}\right)$$

$$= \left(4\pi r.\Delta\rho_\alpha\right)^3\left[1 + \frac{1}{\Delta\rho_\alpha}\left(\rho_{ca} - \rho_w\right)\right]$$

Equation 29

Finally, R may be reduced to the alternative expressions:-

$$R = \left(4\pi r.\Delta\rho_\alpha\right)^3\left(1 + \frac{\Delta\rho_\beta}{\Delta\rho_\alpha}\right) \qquad \textbf{Equation 30a}$$

$$R = \left(4\pi r.\Delta\rho_\alpha\right)^3\left(1 - \frac{\Delta\rho_\gamma}{\Delta\rho_\alpha}\right) \qquad \textbf{Equation 30b}$$

$$R = \left(4\pi r.\Delta\rho_\alpha\right)^3\left(1 - \frac{\rho_w - \rho_{ca}}{\rho_s - \rho_{ca}}\right) \qquad \textbf{Equation 30c}$$

<u>The Mechanical Expression of C</u>

The substitution of Equation Thirty a into Equation Twenty-Three shows that:-

$$C = \sqrt[3]{\left(4\pi r.\Delta\rho_\alpha\right)^3\left(1 + \frac{\Delta\rho_\beta}{\Delta\rho_\alpha}\right)}$$

Equation 31

from which three alternative expressions of C follow:-

$$C = 4\pi r.\Delta\rho_\alpha\sqrt[3]{1 + \frac{\Delta\rho_\beta}{\Delta\rho_\alpha}} \qquad \textbf{Equation 32a}$$

$$C = 4\pi r.\Delta\rho_\alpha\sqrt[3]{1 - \frac{\Delta\rho_\gamma}{\Delta\rho_\alpha}} \qquad \textbf{Equation 32b}$$

$$C = 4\pi r.\Delta\rho_\alpha\sqrt[3]{1 - \frac{\rho_w - \rho_{ca}}{\rho_s - \rho_{ca}}} \qquad \textbf{Equation 32c}$$

PART IV
THE REAL FIRST ROOT AND
THE SOLUTION FOR FLOAT WALL THICKNESS

The First Root, x_1, is the one we want.
x_1 is equivalent to Wall Thickness, t, and is given by:-

$$x_1 = -\frac{k_2}{3k_3} - \frac{C}{3k_3} - \frac{P}{3k_3 C}$$

Equation 33

Substitution of the mechanical equivalents of the Coefficients $k_0,...,k_3$ gives:-

$$x_1 = -\frac{-4\pi r.\Delta\rho_\alpha}{3\left(\dfrac{4}{3}\pi.\Delta\rho_\alpha\right)} - \frac{C}{3\left(\dfrac{4}{3}\pi.\Delta\rho_\alpha\right)} - \frac{P}{3\left(\dfrac{4}{3}\pi.\Delta\rho_\alpha\right).C}$$

Equation 34

Cancellation and elimination of the third term (because P = 0), enables:-

$$x_1 = r - \frac{C}{4\pi.\Delta\rho_\alpha} = r - \frac{\sqrt[3]{R}}{4\pi.\Delta\rho_\alpha}$$

Equation 35

Substitution of Equation Thirty-Two a into Equation Thirty-Five permits:-

$$x_1 = r - \frac{4\pi r.\Delta\rho_\alpha \sqrt[3]{1 + \dfrac{\Delta\rho_\beta}{\Delta\rho_\alpha}}}{4\pi.\Delta\rho_\alpha}$$

Equation 36

Cancellation and re-arrangement of Equation Thirty-Six leads to the formation of these three alternative forms of the first root:-

$$x_1 = r\left(1 - \sqrt[3]{1 + \frac{\Delta\rho_\beta}{\Delta\rho_\alpha}}\right) \qquad \textbf{Equation 37a}$$

$$x_1 = r\left(1 - \sqrt[3]{1 - \frac{\Delta\rho_\gamma}{\Delta\rho_\alpha}}\right) \qquad \textbf{Equation 37b}$$

$$x_1 = r\left(1 - \sqrt[3]{1 - \frac{\rho_w - \rho_{ca}}{\rho_s - \rho_{ca}}}\right) \qquad \textbf{Equation 37c}$$

In terms of Wall Thickness, t,:-

$$t = r\left(1 - \sqrt[3]{1 - \frac{\rho_w - \rho_{ca}}{\rho_s - \rho_{ca}}}\right)$$

Equation 38

Equation Thirty-Eight makes manifest that:-

(i)   When the Float Material and Water Densities are the same, there is no Included Chamber (i.e. the float is solid).

(ii)   As long as the Float Material Density exceeds that of Stream Water: t is less than r, and there is a chamber

(iii)   When $\rho_{ca}=0$ the chamber is a vacuum and a hollow float is still viable as long as $\rho_w<\rho_s$.

(iv)   The chamber can contain a variety of dense substances so long that $\Delta\rho_\gamma/\Delta\rho_\alpha<1$.

For our data the value of Wall Thickness $t\equiv x_1$, is near to 0.007003201859349 meters, or around 7mm, a value I have attempted proportionately to reflect in Figure One.

To check the accuracy of this value it is convenient to back-substitute t into the following equation for Float Mass, M_{float}:-

$$M_{float} = \rho_s\left\{\frac{4}{3}\pi\left[r^3 - (r-t)^3\right] + \rho_{ca}\frac{4}{3}\pi(r-t)^3\right\}$$

Equation 39

M_{float} should prove numerically equal to M_D, the Mass of Displaced Stream Water.

I have a copy of MATHCAD® 6.0 Student Edition for which I took out a £50 licence when I was teaching in the last century. Using the old scratchpad I found that the two masses agreed to at least fifteen places of decimals, the limit of the software's accuracy.

A Simple Inverse Quartic Model
of Stream Cross Section

by
James R Warren BSc MSc PhD PGCE

An old scheme of river course classification divides streams into a proverbial three parts:-

(a)   Immature
Small, high-gradient rivulets that cascade over cataracts and riffles whilst in contact with bedrock, boulders or cobbles.

(b)   Mature
Slower, typically larger rivers of limited sinuosity that make a uniform flow through a bed of incompetent sandy alluvium.

(c)   Old Aged
Sinuous, often tidal, streams that meander over an almost flat expanse of silt or mud.

Examination of literature discloses a typical sectional morphology for mature streams, at least for such as cross the interior plains of North America. This cross-sectional geometry may be symmetrical or skewed toward the left or right bank at places where the course veers.

Where symmetrical, a flat alluvial bed presents between steep banks of matted or semi-consolidated alluvium. Where skewed, a precipitously steep channel is flanked by a long shoulder of detritus at its angle of repose.

To assist other mathematical developments we quest for a tenable mathematical model of stream cross-section that adequately mimics the analytic geometry of the form, both for symmetric and skewed conditions.

It would be convenient if such a model facilitated a finite standardised Stream Width, w_s, on the interval of (say) $0 \le w_s \le 1$. To facilitate Gaussian Quadrature (if desirable) or other mathematical conveniences it would be better if $-1 \le w_s \le +1$.

Perusal of "CRC Standard Curves and Surfaces"[1] offers an algebraic curve function, 2.7.1, that, when inverted, suggests a reasonable model of symmetric mature stream sections seen in literature. The orthostatic version is defined as:-

$$y = \frac{c}{a^4 + x^4}$$

Equation 1

The inverted form is of course:-

$$y = \frac{-c}{a^4 + x^4}$$

Equation 2

We may interpret the Depth Function y as a simple multiple of local stream depth and x as that function's position along the section normal to the flow. y is conventionally negative.

Take y = 0 as the stream air surface.

It is not necessarily the case that Equation Two yields x = 0, x = -1 or x = +1 when y = 0 at the river's banks.

To achieve the conditions (-1,0) and (+1,0) at the air-water-alluvium triple points I had to add a second fractional term as a vertical shift-skew modifier $-c/(a^4+\sigma)$:-

$$y = \frac{-c}{a^4 + \sigma x^4} - \frac{-c}{a^4 + \sigma}$$

Equation 3

where a and σ are arbitrary constants.

Additionally, to model skew morphologies I needed a Skew Function, $e^{\beta x}$, where β is the Skew Factor. β is zero for symmetric sections. For skews of the deep channel to the right bank, β = negative and for left bank skew β = positive.

The composite Stream Section Function can now be specified as:-

$$y = e^{\beta x}\left(\frac{-c}{a^4 + \sigma x^4} - \frac{-c}{a^4 + \sigma}\right)$$

Equation 4

Preliminary studies demonstrated the redundancy of constant a for this application, and I accordingly substituted unity for it. Further, to achieve manageable vertical scaling I also set the scale parameter σ to unity.

Accordingly:-

$$y = -ce^{\beta x}\left(\frac{1}{1 + x^4} - \frac{1}{2}\right)$$

Equation 5

To anchor Equation Five to physical realities it proved necessary, in addition to physical River Width, w_s, to define four boundary measurements:-

(a) x_{max}, d_{max}

The Maximum Channel Depth, d_{max}, and its corresponding Relative Distance, x_{max}, along the section. (x_{max} is negative for negative β).

(b) x_{mid}, d_{mid}

The Channel Depth at the sectional Mid-Point, d_{mid}, and the x value at the mid-stream, i.e. $x_{mid} = 0$.

As aforementioned, there are simple scaling relations between Physical Depths, d, and Model Depths, y; and d and y can be established as numerically equivalent.

<u>The Calculation of Parameters c and β</u>

The Depth Scale Parameter, c, and the Skew Parameter, β, can be computed from knowledge of x_{max}, d_{max} and d_{mid}.

By re-arrangement of Equation Five, and positivising, we can see that:-

$$y = +ce^{\beta x}\left(\frac{1}{1+x^4} - \frac{1}{2}\right)$$

$$= \frac{ce^{\beta x}}{1+x^4} - \frac{ce^{\beta x}}{2}$$

$$= \frac{2ce^{\beta x} - ce^{\beta x}\left(1+x^4\right)}{2\left(1+x^4\right)}$$

Equation 6

If we equivalate d_{mid} and y it is clear that:-

$$d_{mid} = \frac{2ce^{\beta x_{mid}} - ce^{\beta x_{mid}}\left(1+x_{mid}^4\right)}{2\left(1+x_{mid}^4\right)}$$

Equation 7

Now $x_{mid} = 0$. Therefore:-

$$d_{mid} = \frac{2c-c}{2} = \frac{c(2-1)}{2} = \frac{c}{2}$$

Equation 8

Or:-

$$c = 2.d_{mid}$$

Equation 9

The computation of β requires the use of x_{max} and d_{max}.

By back-substitution of c = 2.d_{mid} in Equation Seven note that:-

$$d_{max} = \frac{2\left[2.d_{mid}.e^{\beta x_{max}}\right] - 2.d_{mid}.e^{\beta x_{max}}\left(1 + x_{max}^{4}\right)}{2\left(1 + x_{max}^{4}\right)}$$

$$= 2.d_{mid}.e^{\beta x_{max}}\left[\frac{1}{1 + x_{max}^{4}} - \frac{1}{2}\right]$$

Equation 10

Isolation of multipliers allows us re-arrange Equation Ten as:-

$$\frac{d_{max}}{2d_{mid}e^{\beta x_{max}}} = \frac{1}{1 + x_{max}^{4}} - \frac{1}{2}$$

Equation 11

Taking logarithms:-

$$\ln(d_{max}) - \ln(2) - \ln(d_{mid}) - \beta x_{max} = \ln\left(\frac{1}{1 + x_{max}^{4}} - \frac{1}{2}\right)$$

Equation 12

Or:-

$$\beta = \frac{1}{x_{max}}\left[\ln(d_{max}) - \ln(2) - \ln(d_{mid}) - \ln\left(\frac{1}{1 + x_{max}^{4}} - \frac{1}{2}\right)\right]$$

Equation 13

Knowledge of c and β allows us to compute a synthetic profile as:-

$$v = -ce^{\beta u}\left(\frac{1}{1 + u^{4}} - \frac{1}{2}\right)$$

Equation 14

v permits correlations with standardised measured stream profiles, and also further theoretical developments.

The Length of the Wetted Perimeter

Differentiation of the synthetic Equation Fourteen with respect to u enables us to write:-

$$\frac{dv}{du} = -c\beta e^{\beta u}\left(\frac{1}{1+u^4} - \frac{1}{2}\right) + \frac{4ce^{\beta u}}{\left(1+u^4\right)^2}.u^3$$

$$= -ce^{\beta u}\left[\beta\left(\frac{1}{1+u^4} - \frac{1}{2}\right) - \frac{4u^3}{\left(1+u^4\right)^2}\right]$$

Equation 15

Since u is defined on the interval $-1\leq u\leq +1$, we may write the curve length integral (rectification) of Equation Fifteen as:-

$$S_w = \int_{-1}^{+1}\sqrt{1+\left(\frac{dv}{du}\right)^2}.du$$

$$= \int_{-1}^{+1}\sqrt{1+\left(-ce^{\beta u}\left[\beta\left(\frac{1}{1+u^4} - \frac{1}{2}\right) - \frac{4u^3}{\left(1+u^4\right)^2}\right]\right)^2}.du$$

Equation 16

Equation Sixteen produces a positive S_w on the interval $2\leq S_w\leq\infty$ irrespective of the sign of the differential.

A discrete segmental approximation of S_w is of course available as:-

$$L_{ap} = \sum_{i=0}^{n}\sqrt{\left(u_{i+1} - u_i\right)^2 + \left(v_{i+1} - v_i\right)^2}$$

Equation 17

For arbitrary n.

For $n = 48$, $c = 0.67$ and $\beta = +1.37$ I computed the relevant Specific Defect as:-

$$SpDef = 100\left(\frac{S_w - L_{ap}}{L_{ap}}\right) = 0.021103511141137\%$$

Equation 18

There is no closed form for the integral of Equation Sixteen.

<u>The Area Integral</u>

The standardised cross-sectional Area of the Water Flow, A_w, is given by:-

$$A_w = \int_{-1}^{+1} -ce^{\beta u}\left(\frac{1}{1+u^4} - \frac{1}{2}\right).du = -c\int_{-1}^{+1} e^{\beta u}\left(\frac{1}{1+u^4} - \frac{1}{2}\right).du$$

Equation 19

There is no closed form for the integral of Equation Nineteen.

During my numerical experiments I used Simpson's One-Third Rule[2] to estimate A_w. I chose this rule because it is stable, adequately precise and simply programmed. It is of course unrealistic to the extent that it approximates the envelope as a series of parabolic segments, but higher polynomial fitments would be even less representative of reality, and even more expensive.

In my MATHCAD® jottings I used n = 48 panels and in the EXCEL® elaborations n = 100.

In either case, precision went far beyond the conceivable fit of the geomorphological paradigm invented.

In this standardised model, using i = 0...n, the Panel Width, h, is given by:-

$$h = \frac{u_n - u_0}{n} = \frac{2}{n}$$

Equation 20

where $u_n = +1$ and $u_0 = -1$.
Also:-

$$u_i = u_0 + h.i$$

Equation 21

The requisite terms of the Simpson's Rule integration are:-

$$g_0 = e^{-1\beta}.\left[\frac{1}{1+(-1)^4} - \frac{1}{2}\right] = 0$$

Equation 22a

$$g_i = e^{\beta(u_0+ih)}.\left[\frac{1}{1+(u_0+ih)^4} - \frac{1}{2}\right]$$

Equation 22b

and:-

$$g_n = e^{1\beta} \cdot \left[\frac{1}{1+1^4} - \frac{1}{2}\right] = 0$$

Equation 22c

Together the Equations Twenty-Two define a composite Simpson's One-Third Rule Integration as:-

$$A_w = -c \cdot \left[\frac{h}{3}\left(g_0 + g_n + \sum_{j=1}^{n-1}\left\{3-(-1)^j\right\} \cdot g_j\right)\right]$$

Equation 23

This may be arranged as:-

$$A_w = \frac{-c.h}{3}\left[\sum_{j=1}^{n-1}\left\{3-(-1)^j\right\} \cdot \left\{e^{\beta(u_0+jh)} \cdot \left[\frac{1}{1+(u_0+jh)^4} - \frac{1}{2}\right]\right\}\right]$$

Equation 24

This yields A_w as conventionally negative, but its absolute value $|A_w|$ is the Standardised Flow Cross-Sectional Area.

For adequately-surveyed stream sections with equidistant plumb points we may specify the trapezoidal Area approximation:-

$$A_{ap} \approx \frac{2}{n}\sum_{i=0}^{n-1}\left[\frac{u_{i+1}+u_i}{2}\right] \approx \frac{2}{n}\sum_{i=0}^{n-1}u_i$$

Equation 25

For $n = 48$, $c = +0.67$ and $\beta = +1.37$ the Specific Defect between the Simpson and Trapezoidal Areas is:-

$$SpDef = 100\left(\frac{A_w - A_{ap}}{A_{ap}}\right) = 0.068291787431961\%$$

Equation 26

<u>The Stream Hydraulic Radius</u>

The Hydraulic Radius, R_h, is the dividend of Sectional Flow Area by Wetted Perimeter:-

$$R_h = \frac{A_w}{S_w}$$

Equation 27

For the Inverse Quartic Model (IQM) this may be specified as:-

$$R_h = \frac{-c\int_{-1}^{+1} e^{\beta u}\left(\frac{1}{1+u^4} - \frac{1}{2}\right).du}{\int_{-1}^{+1}\sqrt{1+\left(+ce^{\beta u}\left[\beta\left(\frac{1}{1+u^4} - \frac{1}{2}\right) - \frac{4u^3}{(1+u^4)^2}\right]\right)^2}.du}$$

Equation 28

Or in terms of the Composite Simpson's One-Third Rule for the Area:-

$$R_h = \frac{\dfrac{ch}{3}\left[\displaystyle\sum_{j=1}^{n-1}\left\{3-(-1)^j\right\}.\left\{e^{\beta(u_0+jh)}.\left[\frac{1}{1+(u_0+jh)^4} - \frac{1}{2}\right]\right\}\right]}{\int_{-1}^{+1}\sqrt{1+\left(+ce^{\beta u}\left[\beta\left(\frac{1}{1+u^4} - \frac{1}{2}\right) - \frac{4u^3}{(1+u^4)^2}\right]\right)^2}.du}$$

Equation 29

For n = 48, c = 0.67 and β = +1.37 the Hydraulic Radius is 0.242912474193561 using MATHCAD® intrinsic integration, or 0.242912433613257 using Simpson's Rule.

Metric Limits

It should be noted that parameters c and β are not restricted to the values shown. Indeed in this standardised model the following limits of the constants, variables and parameters of The Inverse Quartic Model may theoretically be specified as:-

(a) $\quad -1 \quad \leq u \quad \leq +1$
(b) $\quad 0 \quad \leq v \quad \leq -\infty$
(c) $\quad 0 \quad \leq c \quad \leq +\infty$
(d) $\quad -\infty \quad \leq \beta \quad \leq +\infty$
(e) $\quad 2 \quad \leq S_w \quad \leq +\infty$
(f) $\quad 0 \quad \leq A_w \quad \leq +\infty$

In the stream sections context it should be noted that any value exceeding ±2 (or +3 in the case of perimeter S_w) would be exceptional.

<u>Illustrative Diagrams</u>

Figure One is a two-dimensional projection in elevation of five IQM stream sections for c = 0.5 and β = {-1,-0.5, 0, +0.5, +1}.

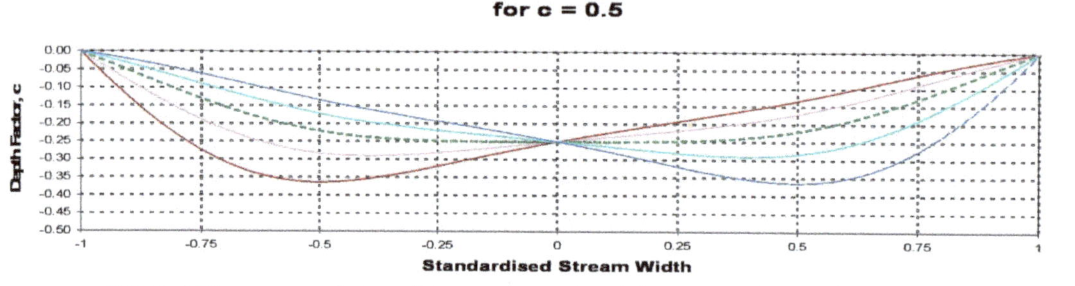

Figure 1
Some IQM Specimen Sections

It is clear that the Inverse Quartic Model permits a crick discontinuity to manifest between quasi-linear bend-bar slopes at a sedimentary angle of repose; and the incised outer channel. This more or less mirrors documented stream sections from a variety of mature river studies and even mimics certain open channels over ice.

Figures Two to Six illustrate, as three-dimensionally rendered ribbons, the sections for c = {0, 0.25, 0.5, 0.75, 1.0} and β = {-1,-0.5, 0, +0.5, +1}.

Clearly, for c = 0 the "section" is a two-unit line across the flow surface; Hence S_w = 2 and A_w = 0.

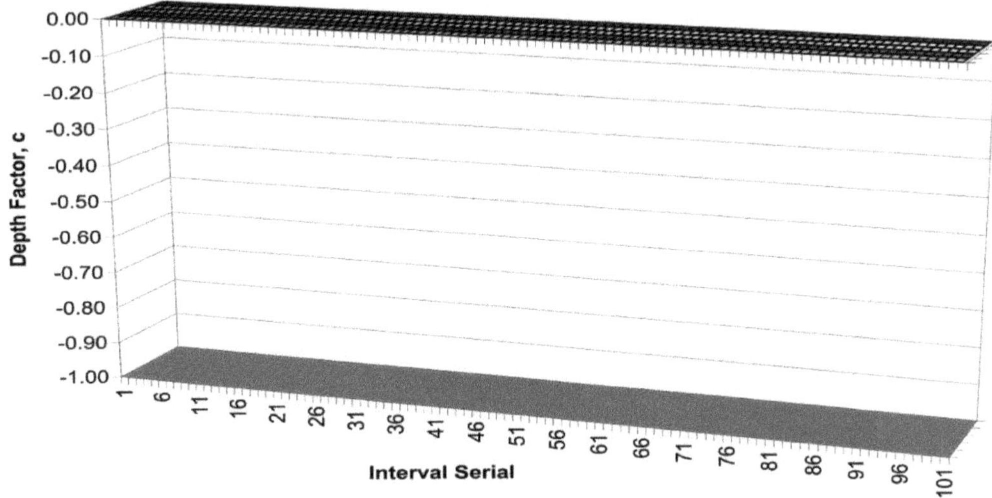

Figure 2
IQM Sections for Depth Parameter c = 0
$\beta = \{-1, -0.5, 0, +0.5, +1\}$

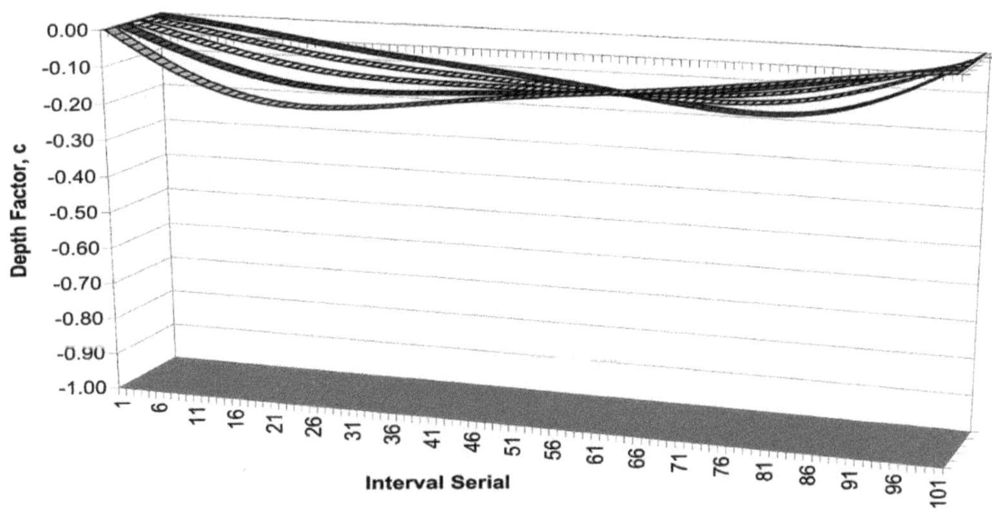

Figure 3
IQM Sections for Depth Parameter c = 0.25
$\beta = \{-1, -0.5, 0, +0.5, +1\}$

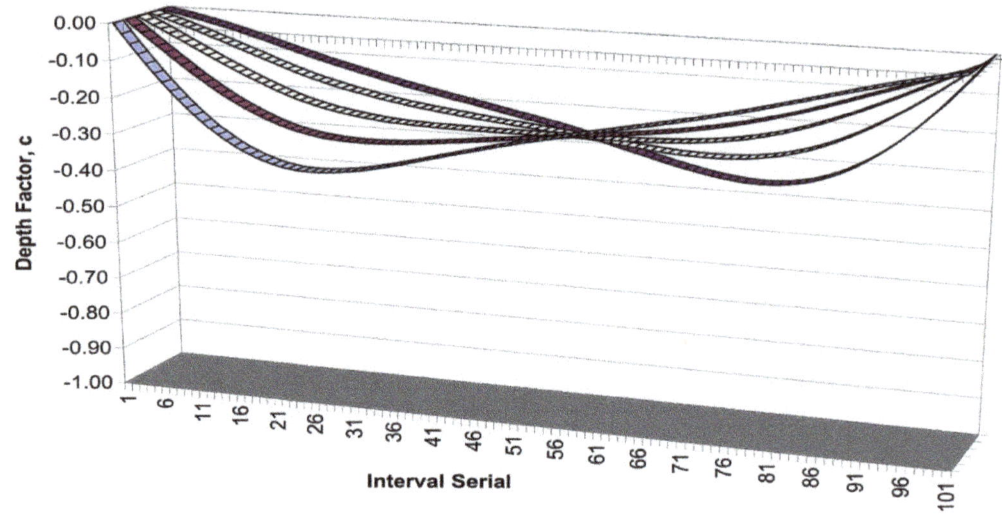

Figure 4
IQM Sections for Depth Parameter c = 0.5
$\beta = \{-1, -0.5, 0, +0.5, +1\}$

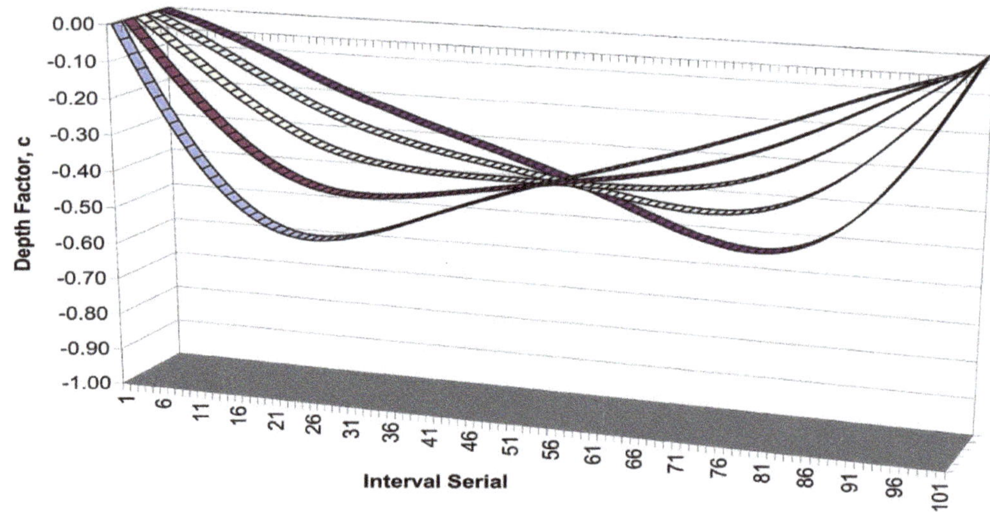

Figure 5
IQM Sections for Depth Parameter c = 0.75
$\beta = \{-1, -0.5, 0, +0.5, +1\}$

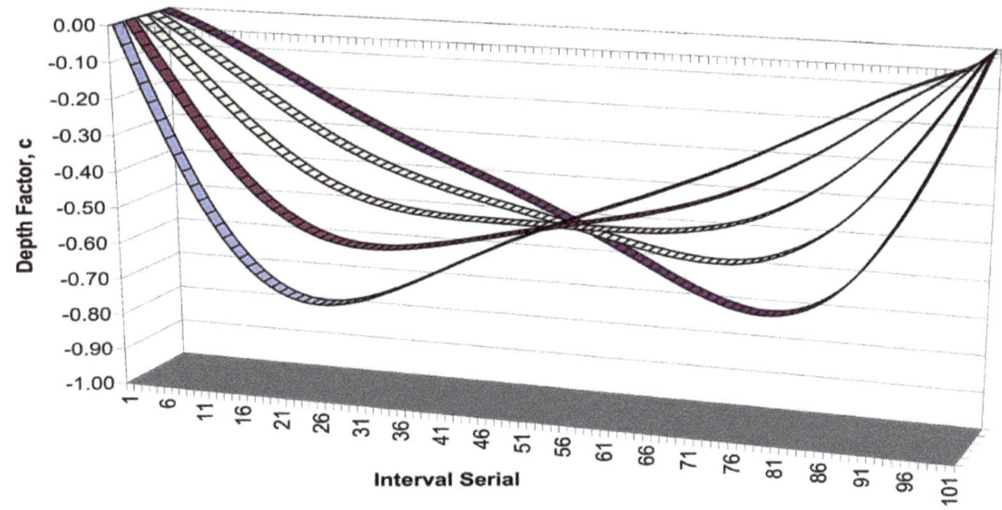

Figure 6
IQM Sections for Depth Parameter c = 1.0

$\beta = \{-1, -0.5, 0, +0.5, +1\}$

<u>The Structures of Output Metrics</u>

For the given values of c and β, Tables One, Two and Three respectively tabulate values of A_w, S_w and R_h.

	β				
	-1	**-0.5**	**0**	**0.5**	**1**
c **0**	0.000000	0.000000	0.000000	0.000000	0.000000
0.25	0.203458	0.188347	0.183486	0.188347	0.203458
0.5	0.406915	0.376695	0.366973	0.376695	0.406915
0.75	0.610373	0.565042	0.550459	0.565042	0.610373
1	0.813830	0.753389	0.733946	0.753389	0.813830

Table 1
Selected Values of IQM
Sectional Area

	β				
	-1	**-0.5**	**0**	**0.5**	**1**
c **0**	2.000000	2.000000	2.000000	2.000000	2.000000
0.25	2.052764	2.032000	2.026082	2.032000	2.052764
0.5	2.188387	2.120767	2.100443	2.120767	2.188387
0.75	2.375964	2.252230	2.214018	2.252230	2.375964
1	2.598239	2.414425	2.357072	2.414425	2.598239

Table 2
Selected Values of IQM
Wetted Perimeter

	β				
	-1	**-0.5**	**0**	**0.5**	**1**
c **0**	0.000000	0.000000	0.000000	0.000000	0.000000
0.25	0.099114	0.092691	0.090562	0.092691	0.099114
0.5	0.185943	0.177622	0.174712	0.177622	0.185943
0.75	0.256895	0.250881	0.248625	0.250881	0.256895
1	0.313224	0.312037	0.311380	0.312037	0.313224

Table 3
Selected Values of IQM
Hydraulic Radius

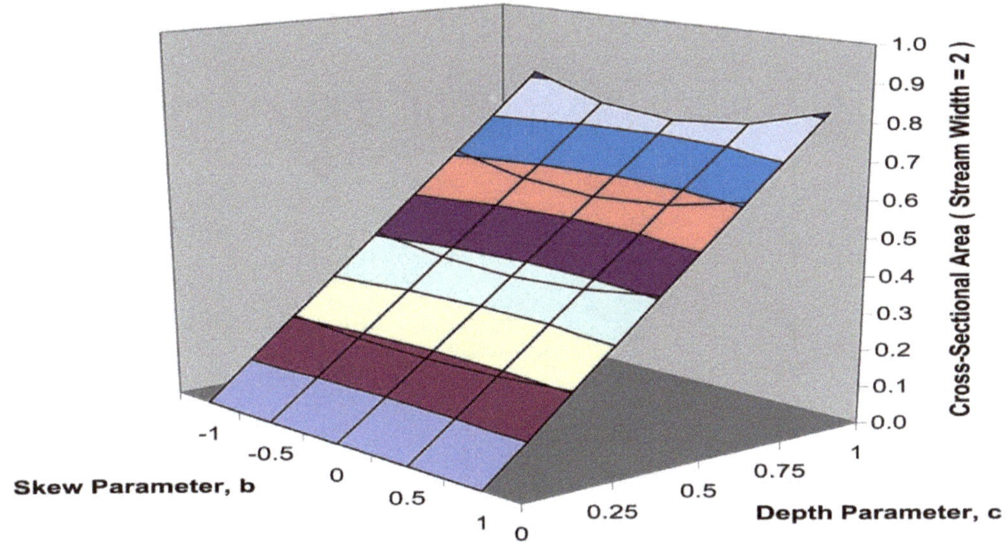

Figure 7
IQM Parametric Relations for Sectional Area

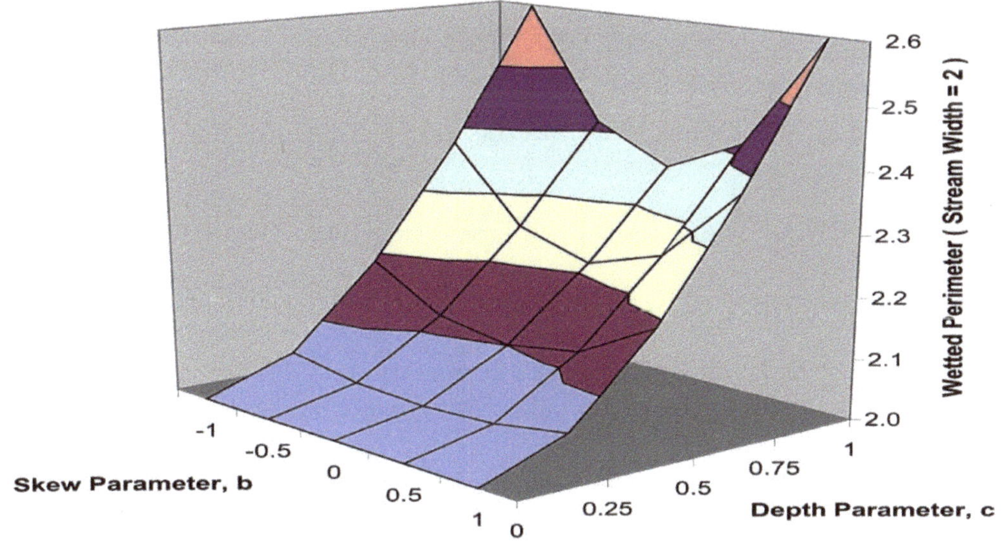

Figure 8
IQM Parametric Relations for Wetted Perimeter

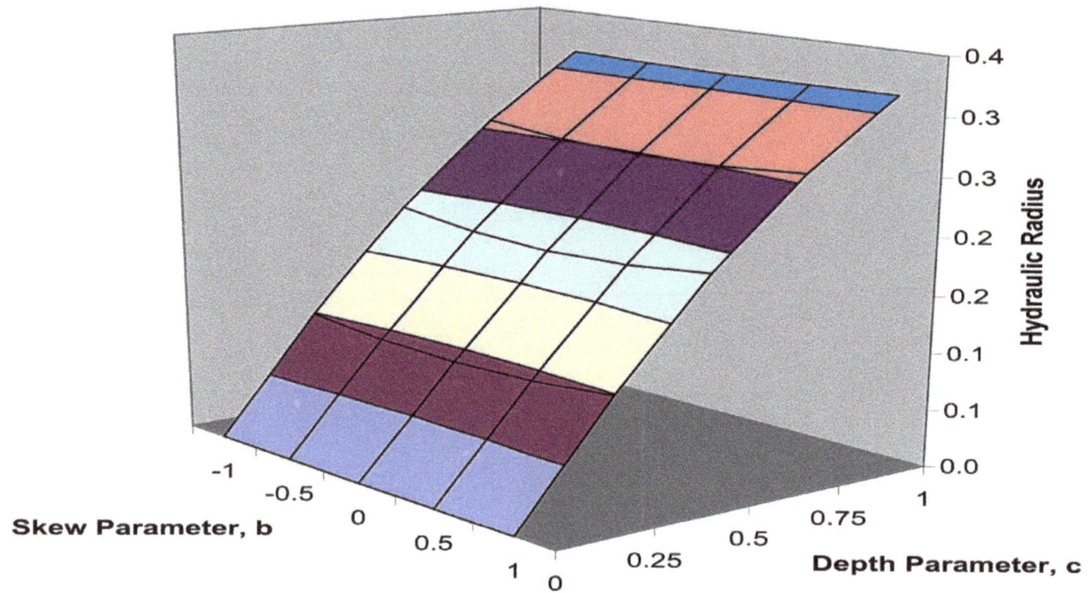

Figure 9
IQM Parametric Relations for Hydraulic Radius

The geometries displayed are symmetrical and highly-suggestive of the viability of further simplifications.

<u>The Failure of Elementary Methods of Integration</u>

In order to explore each avenue that I knew I attempted to apply Integration by Parts and Integration by Substitution to the Stream Section Area Equation:-

$$A_w = -c \int_{-1}^{+1} e^{\beta x} \left(\frac{1}{1+x^4} - \frac{1}{2} \right) . dx$$

Equation 30

Both methods failed to produce a correct value.

An essential property of the failure of such techniques is the singular discontinuity of the integrand at symmetry, i.e. $\beta = 0$.

This is illustrated by the resolution of the subtrahend developed below:-

$$A_w = -c \int_{-1}^{+1} e^{\beta x} \left(\frac{1}{1+x^4} - \frac{1}{2} \right) . dx$$

$$= -c \int_{-1}^{+1} \left(\frac{e^{\beta x}}{1+x^4} - \frac{e^{\beta x}}{2} \right) . dx$$

$$= -c \int_{-1}^{+1} \left(\frac{e^{\beta x}}{1+x^4} \right) . dx - -c \int_{-1}^{+1} \left(\frac{e^{\beta x}}{2} \right) . dx$$

Equation 31

The singularity manifests when we integrate the subtrahend:-

$$-c \int_{-1}^{+1} \frac{e^{\beta x}}{2} . dx = -c \left[\frac{e^{\beta x}}{2\beta} \right]_{-1}^{+1}$$

Singular Expression 1

Any attempt to shift the origin of the Skew Factor will of course only complicate matters by introducing problems of its own.

When I offered up the minuend of the integrand of Equation Thirty to the Wolfram Mathematica® integration engine[3] the program delivered the result:-

$$\int_{-1}^{+1} \frac{-ce^{\beta x}}{1+x^4} . dx = -\frac{1}{4} c \, \mathrm{Rootsum} \left[\#1^4 + 1 \&, \frac{e^{\#1.\beta} . Ei(\beta x - \beta \#1)}{\#1^3} \& \right]$$

Equation 32

Ei(z) is the Exponential Integral of its argument z defined by:-

$$Ei(z) = \gamma + \ln(z) + \sum_{i=1}^{n} \frac{z^i}{n \times n!}$$

Equation 33

where γ is the Euler-Mascheroni Constant. ($\gamma \approx 0.577215664901533$).

The integrals that I managed to compute using the Equation Thirty-Two expression were both complex and numerically-incorrect. Quite possibly I mis-applied the Mathematica® result, but I was satisfied that any method that appealed to exotic functions would be prohibitively expensive for use in environmental geoscience where of course the quality of the data seldom justifies extreme sophistication.

<u>Gaussian-type Quadrature Applications</u>

Gaussian-type quadrature techniques are schemes of numerical integration that seek to generate high-accuracy numerical results using *unevenly* spaced abscissas: That is panels of unequal width. These abscissas are not of course random but judiciously-generated using a mathematical process designed to give exact integrations for certain families of integrand algebras.

In contrast, Simpson's Rule and other Closed-Form Newton-Cotes schemes use *evenly* spaced abscissas that fit polynomials between data points whether the said polynomials are appropriate or not.

Gauss-Legendre Quadrature (GLQ)

Gauss-Legendre Quadrature, sometimes simply called Gaussian Quadrature, is the original of these techniques and is exact for the family of integrand objects known as Legendre Polynomials. Gaussian Quadrature is about right for most integrands featuring algebraic polynomials and I found that Gauss-Legendre Quadrature on thirteen points was roughly good for Stream Section Area.

GLQ applications are predicated upon knowledge of the Weight, w_0, and also the i = 0...n Abscissas, x_i, and Weights, w_i. These statistics are computed using specialised functions and extensively-published[4].

The relevant thirteen $\{w_0, (x_i, w_i)\}$ for my trials are tabulated below:-

i	Abscissor, x_i	Weight, w_i
0		0.232551553230874
1	0.230458315955135	0.226283180262897
2	0.448492751036447	0.207816047536889
3	0.642349339440340	0.178145980761946
4	0.801578090733310	0.138873510219787
5	0.917598399222978	0.092121499837728
6	0.984183054718588	0.040484004765316

Table 4
Gauss-Legendre Abscissas and Weights
for n = 13

If we define the (positivised) integrand as:-

$$f(x) = e^{\beta x}\left(\frac{1}{1+x^4} - \frac{1}{2}\right)$$

Equation 34

and the predicates:-

$$n = 13 \qquad\qquad m = \text{entier}\left(\frac{n}{2}\right)$$

Equation 35a   **Equation 35b**

then the appropriate Gauss-Legendre Integral is:-

$$A_{GL} = c \left[\sum_{i=1}^{m} w_i . f(-x_i) + \sum_{i=1}^{m} w_i . f(x_i) + \frac{w_0}{2} \right]$$

Equation 36

As with the Simpson versus Trapezoidal trial above we will document Gaussian-type trials for $c = +0.67$ and $\beta = +1.37$.

For $n = 13$ and these parameters, A_{GL} was 0.59539922178237. The MATHCAD® intrinsic integration, A, of Equation Thirty gave 0.595399515241958. Accordingly, the Specific Defect for $(A_{GL}-A)/A$ was -0.000049287844661%.

For a variety of extreme and moderate c and β values specific defect was found to remain less than plus or minus one percent *for the case of areal integration.*

The process was repeated for the curve length integral (wetted perimeter) defined by the differential of the Equation Thirty integrand (with constant c incorporated):-

$$f(x) = \sqrt{1 + \left[ce^{\beta x} \left[\beta \left(\frac{1}{1+x^4} - \frac{1}{2} \right) - \frac{4x^3}{\left(1+x^4\right)^2} \right] \right]^2}$$

Equation 37

For this case, the MATHCAD® intrinsic was 2.451086619650186 and the GLQ approximation 2.31152099922468, yielding a specific defect of -5.6940305294239335%.

The wetted perimeter GLQ defect was found to lie between $\pm1\%$ and $\pm10\%$ for a variety of c and β: Values unacceptable to me, so the hunt was on for a cheap and accurate method for both A_w and P_w.

Chebyshev-Gauss Quadrature

Chebyshev-Gauss Quadrature[5] (CGQ) is optimised for a family of integrands involving the reciprocal square root of an algebraic polynomial.

An attraction of Chebyshev-Gauss Quadrature is that its Abscissas and Weights can be computed using very simple functions.

The Number of Abscissas of Integration, n, can be defined arbitrarily. My ancient MATHCAD® Student Edition package limits me to an array of fifty elements for $i = 0...49$. Accordingly, I defined $n = 48$ and $k = 1...n$.

All of the Weights, w_k, are fixed as simply:-

$$w = w_0 = \frac{\pi}{n+1}$$

Equation 38

whilst the Abscissas, ξ_k, are given by:-

$$\xi_k = \cos\left[\frac{\left(k - \frac{1}{2}\right).\pi}{n+1}\right]$$

Equation 39

In addition:-

$$g(\xi_k) = \sqrt{1 - \xi_k^2} \cdot f(\xi_k)$$

Equation 40

Taking f(x) as the integrand Equation Thirty-Four we may write the Chebyshev-Gauss Integral, A_{CG}, as:-

$$A_{CG} = c.\sum_{k=1}^{n} w_k \cdot g(\xi_k) = c.w\sum_{k=1}^{n} g(\xi_k) = \frac{c.\pi}{n+1} \cdot \sum_{k=1}^{n}\left[\sqrt{1 - \xi_k^2} \cdot f(\xi_k)\right]$$

Equation 41

In its fully-expanded form, A_{CG}, may be written as:-

$$A_{CG} = \frac{c.\pi}{n+1} \cdot \sum_{k=1}^{n}\left[\sqrt{1 - \cos\left[\frac{\left(k - \frac{1}{2}\right).\pi}{n+1}\right]^2} \cdot f\left(\cos\left[\frac{\left(k - \frac{1}{2}\right).\pi}{n+1}\right]\right)\right]$$

Equation 42

Using c = 0.67, β = +1.37 as before, A_{CG} was found to be 0.595398874290055 yielding the slightly-vitiated defect of –0.0001107650726432%.

By applying Chebyshev-Gauss Quadrature to the wetted perimeter integrand Equation Thirty-Seven (with constant c removed) I determined a value for that curve of 2.449660031101819. This leads to a specific defect on the intrinsic S of –0.058202290238578%, which would be in an acceptable ball-park figure for a geomorphological result.

Using these CGQ outcomes, R_h, the Hydraulic Radius, was calculated to be 0.243053675502169, with a specific defect of +0.058128471613913%.

There is currently insufficient evidence to claim that CGQ is superior to 48-panel Simpson's One Third Rule for these parameters.

For a MATHCAD® intrinsic Area, A, and a Simpson's A_{simp} for n = 48, the Specific Defect $(A-A_{simp})/A_{simp}$ proved to be –0.000107650726432%. This is worse than the *area* GLQ result but well within the limits of metrical error.

A Note on Lobatto Quadrature.

For a small number of trial c and β values Gauss-Lobatto Quadrature proved highly accurate for *sectional areas* but unusable for *wetted perimeters*.

<u>References</u>

1   "CRC Standard Curves and Surfaces"
David Henry von Seggern
CRC Press Incorporated of Boca Raton   1993
ISBN 0-8493-0196-3
pp388

      (Algebraic Curves) Function with a^4+x^4
      2.7.1
      Page 54
      $y = c/(a^4+x^4)$

2   "Teach Yourself Calculus"
P Abbott BA
The English Universities Press of London EC1
1940   (corrected edition 1963)
pp 380

      Curve Length Integral on Page 275

3   WolframAlpha
Integral Calculator: Wolfram Mathematica Online Integrator
http://integrals.wolfram.com

4   Gaussian Quadrature Weights and Abscissae
http://pomax.github.io/bezierinfo/legendre-gauss.html

      (All Gauss-Legendre coefficients for n < 65)

5   Rice University Lecture
CAAM 453/553 – NUMERICAL ANALYSIS I
Lecture 26: More on Gaussian Quadrature
http://www.caam.rice.edu/~embree/caam453/lecture26.pdf

**AN APPROXIMATION OF K(½) FOR THE EVALUATION OF
DEVAUCHELLE'S STREAM SLOPE CONSTANT**

by
James R Warren BSc MSc PhD PGCE

This concerns the accurate approximation of the dimensionless factor:-

$$f_{Dev} = \frac{2\sqrt{2}}{3}.K\left(\frac{1}{2}\right)$$

Equation B1

as it appears in the Stream Slope Relation of Devauchelle et al:-

$$Q_w S^2 = \frac{2\sqrt{2}.K\left(\dfrac{1}{2}\right)}{3.C_f}\sqrt{\theta_t^{\,3}\,g\left(\frac{d_s}{R}\right)^5} \equiv Q_0$$

Equation B2

where f_{Dev} is Devauchelle's Dimensionless Slope Factor; $K(½)$ is the Complete Elliptic Integral of the First Kind (CEIF) of One Half; Q_w is Water Discharge; S is Stream Slope; C_f is (Bed) Friction Factor; θ_t is Threshold Shield's Parameter; g is the Acceleration Due to Gravity; d_s is Mean Sediment Particle Diameter; and R is the Differential Relative Density between Water and Sediment.
R is defined by:-

$$R = \frac{\rho_w}{\rho_s - \rho_w}$$

Equation B3

where ρ_w is the Density of Water and ρ_s is the Sediment Density.
Devauchelle et al give a value of 1.75 for f_{Dev}.
Our task here is to identify a simple, efficient and high-precision method for f_{Dev} by exploiting the known relation:-

$$K(k) = \frac{\pi}{2.AGM(1,\sin(\alpha))} = \frac{\pi}{2.AGM(1,\sqrt{k})}$$

Equation B4

where:-

$$\alpha = \sin^{-1}\left(\sqrt{k}\right)$$

Equation B5

K is the Complete Elliptic Integral of the First Kind for Argument k where $0<k\leq1$.
π is the Ludolphine Constant and AGM(x,y) is the Arithmetic-Geometric Mean of x and y defined by:-

$$\binom{a_0}{b_0} = \binom{x}{y} \qquad \binom{a_{i+1}}{b_{i+1}} = \left(\begin{array}{c} \dfrac{a_i + b_i}{2} \\ \sqrt{a_i b_i} \end{array}\right) \qquad AGM(x,y) = \left(\begin{array}{c} \dfrac{x + y}{2} \\ \sqrt{xy} \end{array}\right)$$

Equation B6a   **Equation B6b**   **Equation B6c**

whilst:-

$$c_i = \frac{a_i + b_i}{2}$$

Equation B6d

This iteration will yield fifteen-figure accuracy in four passes when presented with the initial conditions:-

$$a_0 = 1 \qquad\qquad b_0 = \sin(\alpha) \qquad\qquad c_0 = \cos(\alpha)$$

Equation B7a   **Equation B7b**   **Equation B7c**

Approximation of the AGM

A three-pass AGM can achieve a precision sufficiently high for many purposes and I dare say for all environmental science applications.

By internal substitutions it is possible to express a three-pass AGM by the construct:-

$$AGM_{(2)} = \frac{\dfrac{\dfrac{\frac{x+y}{2}+\sqrt{xy}}{2} + \sqrt{\frac{x+y}{2}\cdot\sqrt{xy}}}{2} + \sqrt{\dfrac{\frac{x+y}{2}+\sqrt{xy}}{2} + \sqrt{\frac{x+y}{2}\cdot\sqrt{xy}}}}{2}$$

Equation B8

A degree of simplification gives:-

$$AGM_{(2)} = \frac{x}{2^4} + \frac{y}{2^4} + \frac{1}{2^3}\left(x^{\frac{1}{2}}y^{\frac{1}{2}}\right) + \frac{1}{2^{\frac{5}{3}}}\sqrt{x+y}.x^{\frac{1}{4}}y^{\frac{1}{4}} + \frac{1}{2^{\frac{9}{4}}}\sqrt{x+y+2\sqrt{x}\sqrt{y}}.(x+y)^{\frac{1}{4}}.x^{\frac{1}{8}}y^{\frac{1}{8}}$$

Equation B9

and further simplifications bring us to:-

$$AGM_{(2)} = 2^{-\frac{5}{4}}\sqrt{x+y+2\sqrt{xy}}.(x+y)^{\frac{1}{4}}.(xy)^{\frac{1}{8}}$$

Equation B10

For a system where $a_0 = 1$ and $b_0 = \sin(\alpha)$ we may note the identities:-

$$AGM_{(2)} = \frac{\left(x+y+2\sqrt{xy}\right)^{\frac{1}{2}}(x+y)^{\frac{1}{4}}.(xy)^{\frac{1}{8}}}{\left(\sqrt{32}\right)^{\frac{1}{2}}}$$

$$= \frac{\left(1+\sin(\alpha)+2\sqrt{\sin(\alpha)}\right)^{\frac{1}{2}}(1+\sin(\alpha))^{\frac{1}{4}}.(\sin(\alpha))^{\frac{1}{8}}}{\left(\sqrt{32}\right)^{\frac{1}{2}}}$$

$$= 2^{-\frac{5}{4}}(1+\sin(\alpha))\sqrt{\sqrt{1+\sin(\alpha)}.\sqrt[4]{\sin(\alpha)}}$$

Equation B11

Any of these identities reduce to:-

$$AGM_{(2)} = \frac{(1+\sin(\alpha))^{\frac{1}{4}}.\left[(\sin(\alpha))^{\frac{1}{8}} + (\sin(\alpha))^{\frac{5}{8}}\right]}{\sqrt[4]{32}}$$

Equation B12

The Specific Defect, SpDef$_{AGM-PA}$, is given by:-

$$SpDef_{AGM-Eqn.B12} = 100\frac{Eqn.B12 - AGM}{AGM}$$

Equation B13

For the case of:-

$$\sin(\alpha) = \sqrt{\frac{1}{2}} = 0.707106781186548 \equiv \cos(\alpha)$$

the Specific Defect vis-a-vis the full AGM is -0.000000000048646, or 12-figure accuracy for the pseudoanalytic form.

CEIF(½) and the Devauchelle Coefficient

By employing Equation B12 we may now frame a high-precision estimate of Devauchelle's Stream Slope Coefficient:-

$$f_{Dev} = \frac{2\sqrt{2}}{3}.K\left(\frac{1}{2}\right) = \frac{2\sqrt{2}}{3}.\frac{\pi}{2.AGM(1,\sin(\alpha))}$$

Equation B14

Accordingly:-

$$f_{Dev} = \frac{2\sqrt{2}}{3}.K\left(\frac{1}{2}\right) = \frac{2\sqrt{2}}{3}.\frac{\pi}{2.\dfrac{(1+\sin(\alpha))^{\frac{1}{4}}.\left[(\sin(\alpha))^{\frac{1}{8}}+(\sin(\alpha))^{\frac{5}{8}}\right]}{\sqrt[4]{32}}}$$

Equation B15

Equation B15 may be simplified in the following terms:-

$$f_{Dev} = \frac{2\sqrt{2}}{3} \cdot \frac{2^{\frac{1}{4}}\pi}{(1+\sin(\alpha))^{\frac{1}{4}} \cdot \left[(\sin(\alpha))^{\frac{1}{8}} + (\sin(\alpha))^{\frac{5}{8}}\right]}$$

$$= \frac{2^{\frac{7}{4}}\pi}{3(1+\sin(\alpha))^{\frac{1}{4}} \cdot \left[(\sin(\alpha))^{\frac{1}{8}} + (\sin(\alpha))^{\frac{5}{8}}\right]}$$

$$= \frac{3^{1.10412706875005}\pi}{3(1+\sin(\alpha))^{\frac{1}{4}} \cdot \left[(\sin(\alpha))^{\frac{1}{8}} + (\sin(\alpha))^{\frac{5}{8}}\right]}$$

$$= \frac{1.121195220338285\pi}{(1+\sqrt{k})^{\frac{1}{4}} \cdot \left[k^{\frac{1}{16}} + k^{\frac{5}{16}}\right]}$$

Equation B16

Equation B16 evaluates as 1.748038369613113.

Six-figure accuracy may be tenable to give 1.74804 as the high-precision estimate for Devauchelle's Stream Slope Constant.

TECHNICAL APPENDIX C

**A SIMPLE INVERSE QUARTIC MODEL
OF STREAM CROSS SECTION**

by
James R Warren BSc MSc PhD PGCE

An old scheme of river course classification divides streams into a proverbial three parts:-

    (a)   Immature

Small, high-gradient rivulets that cascade over cataracts and riffles whilst in contact with bedrock, boulders or cobbles.

    (b)   Mature

Slower, typically larger rivers of limited sinuosity that make a uniform flow through a bed of incompetent sandy alluvium.

    (c)   Old Aged

Sinuous, often tidal, streams that meander over an almost flat expanse of silt or mud.

Examination of literature discloses a typical sectional morphology for mature streams, at least for such as cross the interior plains of North America. This cross-sectional geometry may be symmetrical or skewed toward the left or right bank at places where the course veers.

Where symmetrical, a flat alluvial bed presents between steep banks of matted or semi-consolidated alluvium. Where skewed, a precipitously steep channel is flanked by a long shoulder of detritus at its angle of repose.

To assist other mathematical developments we quest for a tenable mathematical model of stream cross-section that adequately mimics the analytic geometry of the form, both for symmetric and skewed conditions.

It would be convenient if such a model facilitated a finite standardised Stream Width, w_s, on the interval of (say) $0 \le w_s \le 1$. To facilitate Gaussian Quadrature (if desirable) or other mathematical conveniences it would be even better if $-1 \le w_s \le +1$.

Perusal of "CRC Standard Curves and Surfaces"[C1] offers an algebraic curve function, 2.7.1, that, when inverted, suggests a reasonable model of symmetric mature stream sections seen in literature. The orthostatic version is defined as:-

$$y = \frac{c}{a^4 + x^4}$$

Equation C1

The inverted form is of course:-

$$y = \frac{-c}{a^4 + x^4}$$

Equation C2

We may interpret the Depth Function y as a simple multiple of local stream depth and x as that function's position along the section normal to the flow. y is conventionally negative.

Take $y = 0$ as the stream air surface.

It is not necessarily the case that Equation C2 yields $x = 0$, $x = -1$ or $x = +1$ when $y = 0$ at the river's banks.

To achieve the conditions $(-1,0)$ and $(+1,0)$ at the air-water-alluvium triple points I had to add a second fractional term as a vertical shift-skew modifier $-c/(a^4+\sigma)$:-

$$y = \frac{-c}{a^4 + \sigma x^4} - \frac{-c}{a^4 + \sigma}$$

Equation C3

where a and σ are arbitrary constants.

Additionally, to model skew morphologies I needed a Skew Function, $e^{\beta x}$, where β is the Skew Factor. β is zero for symmetric sections. For skews of the deep channel to the right bank, β is negative and for left bank skew β is positive.

The composite Stream Section Function can now be specified as:-

$$y = e^{\beta x}\left(\frac{-c}{a^4 + \sigma x^4} - \frac{-c}{a^4 + \sigma}\right)$$

Equation C4

Preliminary studies demonstrated the redundancy of constant a for this application, and I accordingly substituted unity for it. Further, to achieve manageable vertical scaling I also set the scale parameter σ to unity.

Accordingly:-

$$y = -ce^{\beta x}\left(\frac{1}{1 + x^4} - \frac{1}{2}\right)$$

Equation C5

To anchor Equation C5 to physical realities it proved necessary, in addition to physical River Width, w_s, to define four boundary measurements:-

(a) x_{max}, d_{max}

The Maximum Channel Depth, d_{max}, and its corresponding Relative Distance, x_{max}, along the section. (x_{max} is negative for negative β).

(b) x_{mid}, d_{mid}

The Channel Depth at the sectional Mid-Point, d_{mid}, and the x value at the mid-stream, i.e. $x_{mid} = 0$.

As aforementioned, there are simple scaling relations between Physical Depths, d, and Model Depths, y; and d and y can be established as numerically equivalent.

The Calculation of Parameters c and β

The Depth Scale Parameter, c, and the Skew Parameter, β, can be computed from knowledge of x_{max}, d_{max} and d_{mid}.

By re-arrangement of Equation C5, and positivising, we can see that:-

$$y = +ce^{\beta x}\left(\frac{1}{1+x^4} - \frac{1}{2}\right)$$

$$= \frac{ce^{\beta x}}{1+x^4} - \frac{ce^{\beta x}}{2}$$

$$= \frac{2ce^{\beta x} - ce^{\beta x}\left(1+x^4\right)}{2\left(1+x^4\right)}$$

Equation C6

If we equivalate d_{mid} and y it is clear that:-

$$d_{mid} = \frac{2ce^{\beta x_{mid}} - ce^{\beta x_{mid}}\left(1 + x_{mid}^4\right)}{2\left(1 + x_{mid}^4\right)}$$

Equation C7

Now $x_{mid} = 0$. Therefore:-

$$d_{mid} = \frac{2c - c}{2} = \frac{c(2-1)}{2} = \frac{c}{2}$$

Equation C8

Or:-

$$c = 2.d_{mid}$$

Equation C9

The computation of β requires the use of x_{max} and d_{max}.

By back-substitution of c = 2.d$_{mid}$ in Equation C7 note that:-

$$d_{max} = \frac{2\left[2.d_{mid}.e^{\beta x_{max}}\right] - 2.d_{mid}.e^{\beta x_{max}}\left(1 + x_{max}{}^4\right)}{2\left(1 + x_{max}{}^4\right)}$$

$$= 2.d_{mid}.e^{\beta x_{max}}\left[\frac{1}{1 + x_{max}{}^4} - \frac{1}{2}\right]$$

Equation C10

Isolation of multipliers allows us re-arrange Equation C10 as:-

$$\frac{d_{max}}{2d_{mid}e^{\beta x_{max}}} = \frac{1}{1 + x_{max}{}^4} - \frac{1}{2}$$

Equation C11

Taking logarithms:-

$$\ln(d_{max}) - \ln(2) - \ln(d_{mid}) - \beta x_{max} = \ln\left(\frac{1}{1 + x_{max}{}^4} - \frac{1}{2}\right)$$

Equation C12

Or:-

$$\beta = \frac{1}{x_{max}}\left[\ln(d_{max}) - \ln(2) - \ln(d_{mid}) - \ln\left(\frac{1}{1 + x_{max}{}^4} - \frac{1}{2}\right)\right]$$

Equation C13

Knowledge of c and β allows us to compute a synthetic profile as:-

$$v = -ce^{\beta u}\left(\frac{1}{1 + u^4} - \frac{1}{2}\right)$$

Equation C14

v permits correlations with standardised measured stream profiles, and also further theoretical developments.

<u>The Length of the Wetted Perimeter</u>

Differentiation of the synthetic Equation C14 with respect to u enables us to write:-

$$\frac{dv}{du} = -c\beta e^{\beta u}\left(\frac{1}{1+u^4} - \frac{1}{2}\right) + \frac{4ce^{\beta u}}{\left(1+u^4\right)^2}.u^3$$

$$= -ce^{\beta u}\left[\beta\left(\frac{1}{1+u^4} - \frac{1}{2}\right) - \frac{4u^3}{\left(1+u^4\right)^2}\right]$$

Equation C15

Since u is defined on the interval $-1 \leq u \leq +1$, we may write the curve length integral (rectification) of Equation C15 as:-

$$S_w = \int_{-1}^{+1}\sqrt{1+\left(\frac{dv}{du}\right)^2}.du$$

$$= \int_{-1}^{+1}\sqrt{1+\left(-ce^{\beta u}\left[\beta\left(\frac{1}{1+u^4} - \frac{1}{2}\right) - \frac{4u^3}{\left(1+u^4\right)^2}\right]\right)^2}.du$$

Equation C16

Equation C16 produces a positive S_w on the interval $2 \leq S_w \leq \infty$ irrespective of the sign of the differential.

A discrete segmental approximation of S_w is of course available as:-

$$L_{ap} = \sum_{i=0}^{n}\sqrt{\left(u_{i+1} - u_i\right)^2 + \left(v_{i+1} - v_i\right)^2}$$

Equation C17

For arbitrary n.

For $n = 48$, $c = 0.67$ and $\beta = +1.37$ I computed the relevant Specific Defect as:-

$$SpDef = 100\left(\frac{S_w - L_{ap}}{L_{ap}}\right) = 0.021103511141137\%$$

Equation C18

There is no closed form for the integral of Equation C16.

<u>The Area Integral</u>

The standardised cross-sectional Area of the Water Flow, A_w, is given by:-

$$A_w = \int_{-1}^{+1} -ce^{\beta u}\left(\frac{1}{1+u^4} - \frac{1}{2}\right).du = -c\int_{-1}^{+1} e^{\beta u}\left(\frac{1}{1+u^4} - \frac{1}{2}\right).du$$

Equation C19

There is no closed form for the integral of Equation C19.

During my numerical experiments I used Simpson's One-Third Rule[C2] to estimate A_w. I chose this rule because it is stable, adequately precise and simply programmed. It is of course unrealistic to the extent that it approximates the envelope as a series of parabolic segments, but higher polynomial fitments would be even less representative of reality, and even more expensive.

In my MATHCAD® jottings I used n = 48 panels and in the EXCEL® elaborations n = 100.

In either case, precision went far beyond the conceivable fit of the geomorphological paradigm invented.

In this standardised model, using i = 0...n, the Panel Width, h, is given by:-

$$h = \frac{u_n - u_0}{n} = \frac{2}{n}$$

Equation C20

where $u_n = +1$ and $u_0 = -1$.

Also:-

$$u_i = u_0 + h.i$$

Equation C21

The requisite terms of the Simpson's Rule integration are:-

$$g_0 = e^{-1\beta} . \left[\frac{1}{1+(-1)^4} - \frac{1}{2}\right] = 0$$

Equation C22a

$$g_i = e^{\beta(u_0 + ih)} . \left[\frac{1}{1+(u_0 + ih)^4} - \frac{1}{2}\right]$$

Equation C22b

and:-

$$g_n = e^{1\beta} . \left[\frac{1}{1+1^4} - \frac{1}{2}\right] = 0$$

Equation C22c

Together the Equations C22 define a composite Simpson's One-Third Rule Integration as:-

$$A_w = -c\frac{h}{3}\left[g_0 + g_n + \sum_{j=1}^{n-1}\left\{3-(-1)^j\right\}.g_j \right]$$

Equation C23

This may be arranged as:-

$$A_w = \frac{-ch}{3}\left[\sum_{j=1}^{n-1}\left\{3-(-1)^j\right\}.\left\{ e^{\beta(u_0+jh)}.\left[\frac{1}{1+(u_0+jh)^4} - \frac{1}{2} \right] \right\} \right]$$

Equation C24

This yields A_w as conventionally negative, but its absolute value $|A_w|$ is the Standardised Flow Cross-Sectional Area.

For adequately-surveyed stream sections with equidistant plumb points we may specify the trapezoidal Area approximation:-

$$A_{ap} \approx \frac{2}{n}\sum_{i=0}^{n-1}\left[\frac{u_{i+1}+u_i}{2} \right] \approx \frac{2}{n}\sum_{i=0}^{n-1}u_i$$

Equation C25

For n = 48, c = +0.67 and β = +1.37 the Specific Defect between the Simpson and Trapezoidal Areas is:-

$$SpDef = 100\left(\frac{A_w - A_{ap}}{A_{ap}} \right) = 0.0682917874319961\%$$

Equation C26

<u>The Stream Hydraulic Radius</u>

The Hydraulic Radius, R_h, is the dividend of Sectional Flow Area by Wetted Perimeter:-

$$R_h = \frac{A_w}{S_w}$$

Equation C27

For the Inverse Quartic Model (IQM) this may be specified as:-

$$R_h = \frac{c \int_{-1}^{+1} e^{\beta u}\left(\dfrac{1}{1+u^4} - \dfrac{1}{2}\right).du}{\int_{-1}^{+1}\sqrt{1 + \left(+ce^{\beta u}\left[\beta\left(\dfrac{1}{1+u^4} - \dfrac{1}{2}\right) - \dfrac{4u^3}{\left(1+u^4\right)^2}\right]\right)^2}\;.du}$$

Equation C28

Or in terms of the Composite Simpson's One-Third Rule for the Area:-

$$R_h = \frac{\dfrac{ch}{3}\left[\displaystyle\sum_{j=1}^{n-1}\{3-(-1)^j\}.\left\{e^{\beta(u_0+jh)}.\left[\dfrac{1}{1+(u_0+jh)^4} - \dfrac{1}{2}\right]\right\}\right]}{\int_{-1}^{+1}\sqrt{1 + \left(+ce^{\beta u}\left[\beta\left(\dfrac{1}{1+u^4} - \dfrac{1}{2}\right) - \dfrac{4u^3}{\left(1+u^4\right)^2}\right]\right)^2}\;.du}$$

Equation C29

For $n = 48$, $c = 0.67$ and $\beta = +1.37$ the Hydraulic Radius is 0.242912474193561 using MATHCAD® intrinsic integration, or 0.242912433613257 using Simpson's Rule.

<u>Metric Limits</u>

It should be noted that parameters c and β are not restricted to the values shown. Indeed in this standardised model the following limits of the constants, variables and parameters of The Inverse Quartic Model may theoretically be specified as:-

(a)	-1	$\leq u$	$\leq +1$
(b)	0	$\leq v$	$\leq -\infty$
(c)	0	$\leq c$	$\leq +\infty$
(d)	$-\infty$	$\leq \beta$	$\leq +\infty$
(e)	2	$\leq S_w$	$\leq +\infty$
(f)	0	$\leq A_w$	$\leq +\infty$

In the stream sections context it should be noted that any value exceeding ±2 (or $+3$ in the case of perimeter S_w) would be exceptional.

<u>Illustrative Diagrams</u>

Figure C1 is a two-dimensional projection in elevation of five IQM stream sections for $c = 0.5$ and $\beta = \{-1,-0.5, 0, +0.5, +1\}$.

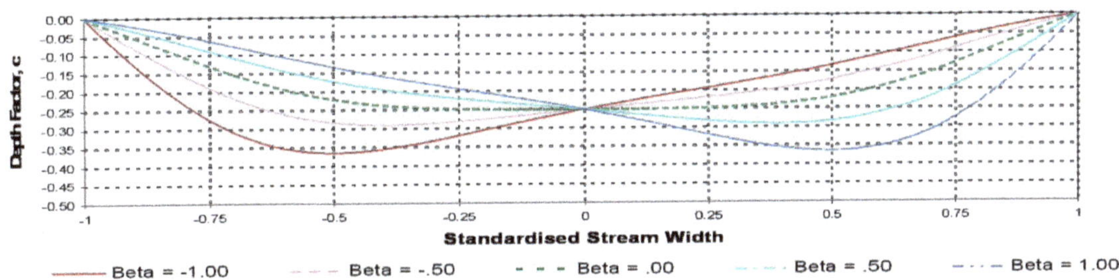

Figure C1
Some IQM Specimen Sections

It is clear that the Inverse Quartic Model permits a crick discontinuity to manifest between quasi-linear bend-bar slopes at a sedimentary angle of repose; and the incised outer channel. This more or less mirrors documented stream sections from a variety of mature river studies and even mimics certain open channels over ice.

Figures C2 to C6 illustrate, as three-dimensionally rendered ribbons, the sections for $c = \{0, 0.25, 0.5, 0.75, 1.0\}$ and $\beta = \{-1, -0.5, 0, +0.5, +1\}$.

Clearly, for $c = 0$ the "section" is a two-unit line across the flow surface; Hence $S_w = 2$ and $A_w = 0$.

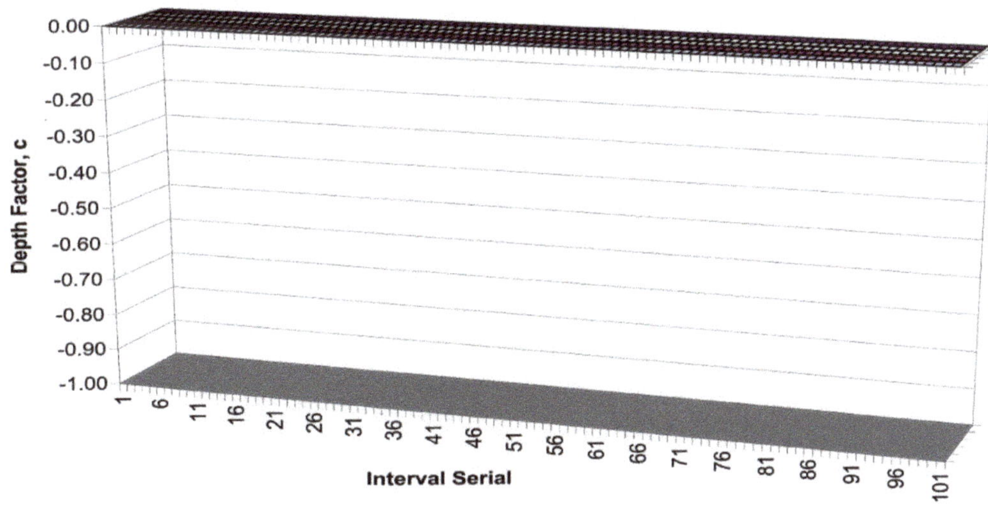

Figure C2
IQM Sections for Depth Parameter $c = 0$
$\beta = \{-1, -0.5, 0, +0.5, +1\}$

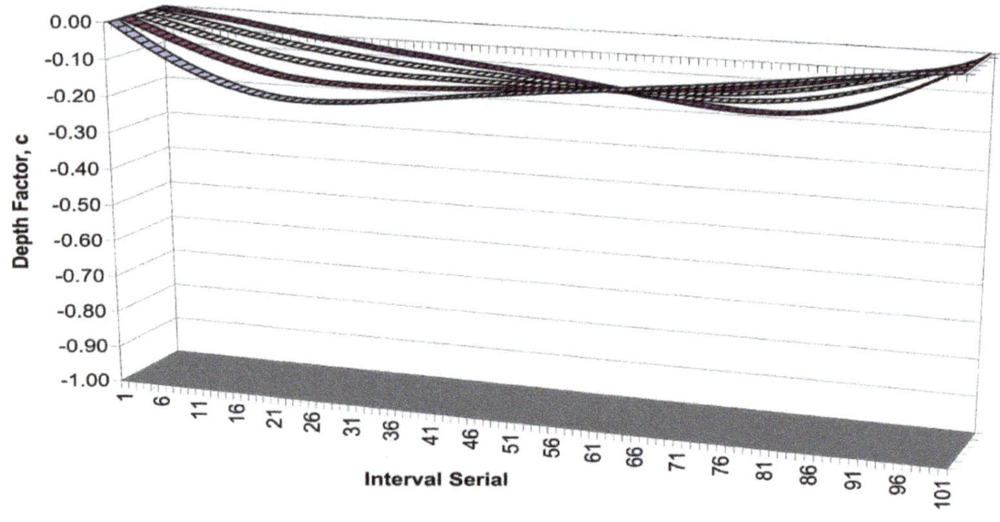

Figure C3
IQM Sections for Depth Parameter c = 0.25
$\beta = \{-1, -0.5, 0, +0.5, +1\}$

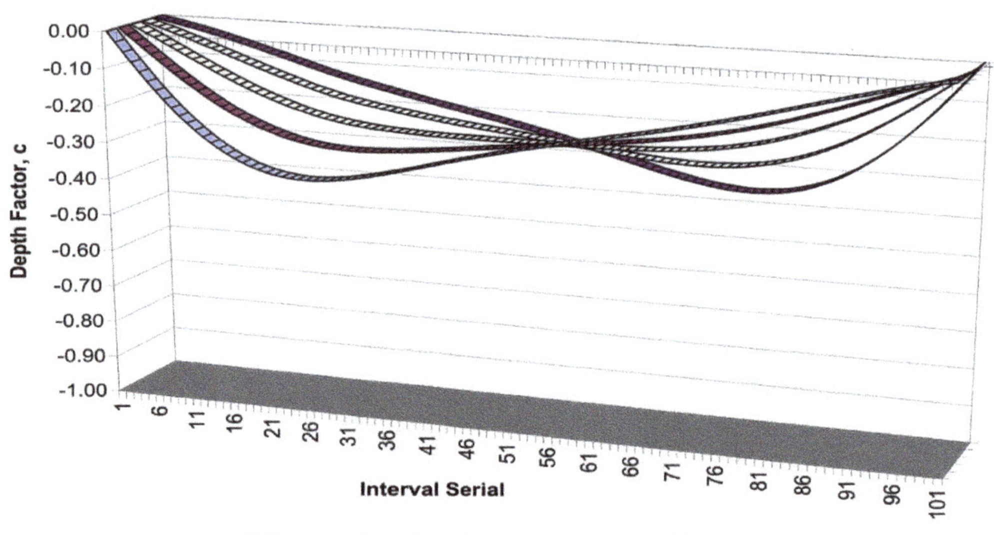

Figure C4
IQM Sections for Depth Parameter c = 0.5
$\beta = \{-1, -0.5, 0, +0.5, +1\}$

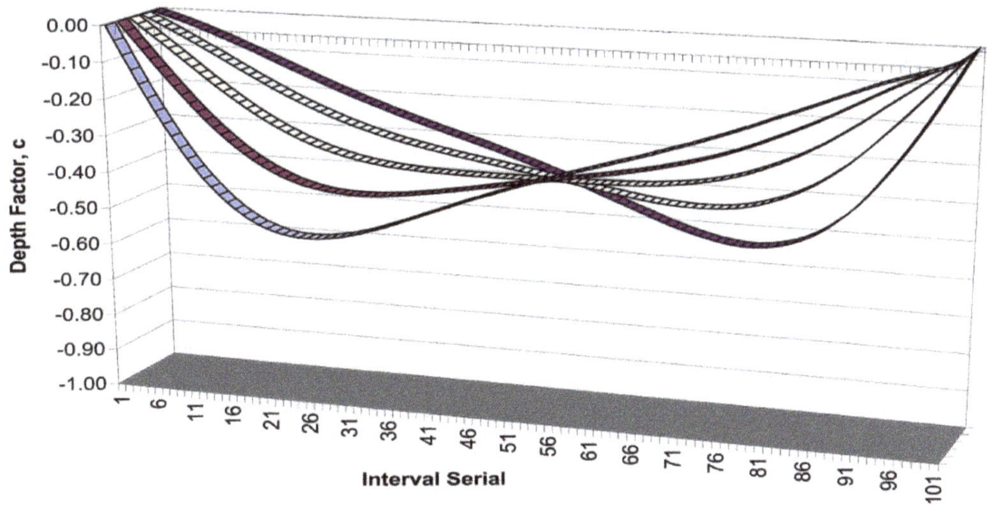

Figure C5
IQM Sections for Depth Parameter c = 0.75
$\beta = \{-1, -0.5, 0, +0.5, +1\}$

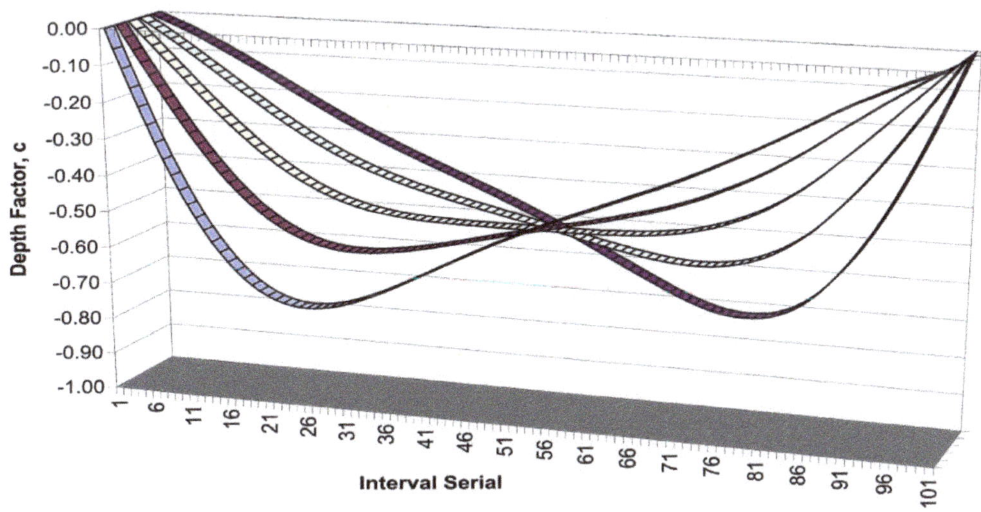

Figure C6
IQM Sections for Depth Parameter c = 1.0

$\beta = \{-1, -0.5, 0, +0.5, +1\}$

For the given values of c and β, Tables C1, C2 and C3 respectively tabulate values of A_w, S_w and R_h.

	β				
	-1	**-0.5**	**0**	**0.5**	**1**
c **0**	0.000000	0.000000	0.000000	0.000000	0.000000
0.25	0.203458	0.188347	0.183486	0.188347	0.203458
0.5	0.406915	0.376695	0.366973	0.376695	0.406915
0.75	0.610373	0.565042	0.550459	0.565042	0.610373
1	0.813830	0.753389	0.733946	0.753389	0.813830

Table C1
Selected Values of IQM
Sectional Area

	β				
	-1	**-0.5**	**0**	**0.5**	**1**
c **0**	2.000000	2.000000	2.000000	2.000000	2.000000
0.25	2.052764	2.032000	2.026082	2.032000	2.052764
0.5	2.188387	2.120767	2.100443	2.120767	2.188387
0.75	2.375964	2.252230	2.214018	2.252230	2.375964
1	2.598239	2.414425	2.357072	2.414425	2.598239

Table C2
Selected Values of IQM
Wetted Perimeter

	β				
	-1	**-0.5**	**0**	**0.5**	**1**
c **0**	0.000000	0.000000	0.000000	0.000000	0.000000
0.25	0.099114	0.092691	0.090562	0.092691	0.099114
0.5	0.185943	0.177622	0.174712	0.177622	0.185943
0.75	0.256895	0.250881	0.248625	0.250881	0.256895
1	0.313224	0.312037	0.311380	0.312037	0.313224

Table C3
Selected Values of IQM
Hydraulic Radius

These relations are respectively illustrated by the plots:-

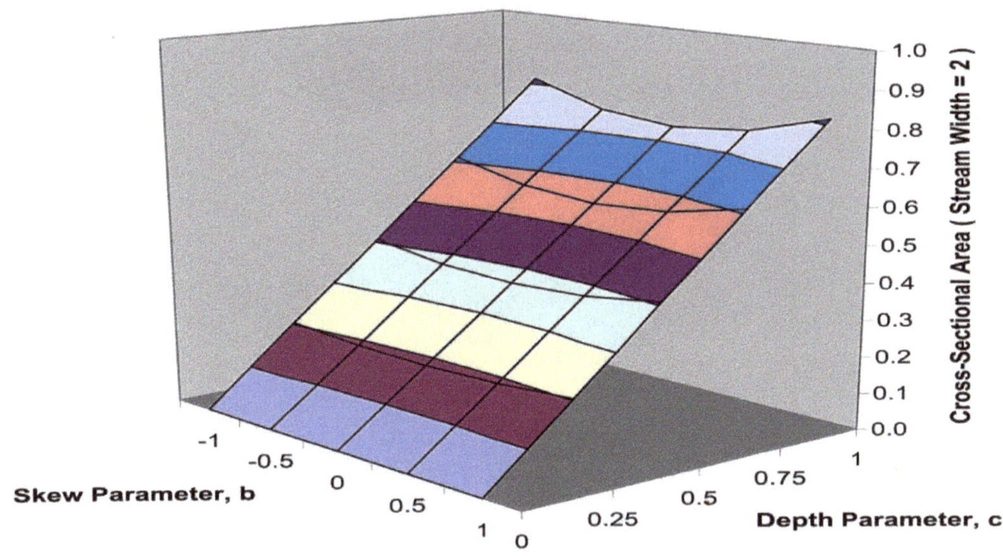

Figure C7
IQM Parametric Relations for Sectional Area

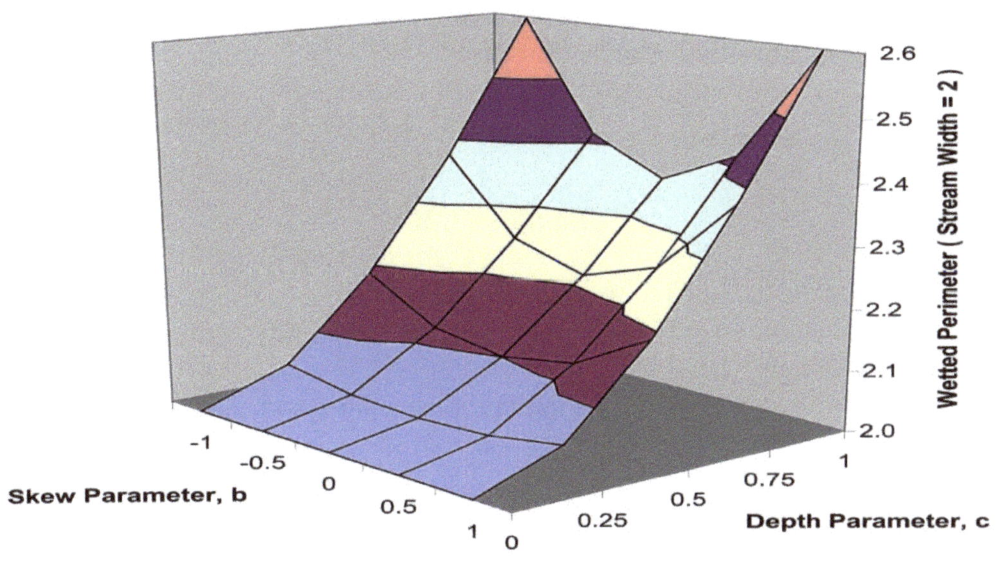

Figure C8
IQM Parametric Relations for Wetted Perimeter

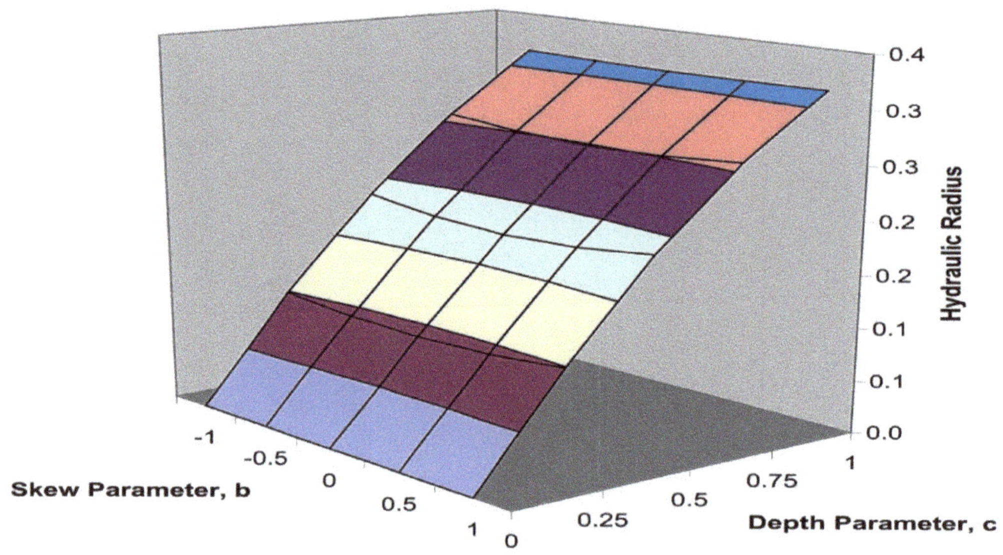

Figure C9
IQM Parametric Relations for Hydraulic Radius

The geometries displayed are symmetrical and highly-suggestive of the viability of further simplifications.

<u>The Failure of Elementary Methods of Integration</u>

In order to explore each avenue that I knew I attempted to apply Integration by Parts and Integration by Substitution to the Stream Section Area Equation:-

$$A_w = -c \int_{-1}^{+1} e^{\beta x} \left(\frac{1}{1+x^4} - \frac{1}{2} \right) . dx$$

Equation C30

Both methods failed to produce a correct value.

An essential property of the failure of such techniques is the singular discontinuity of the integrand at symmetry, i.e. $\beta = 0$.

This is illustrated by the resolution of the subtrahend developed below:-

$$A_w = -c \int_{-1}^{+1} e^{\beta x} \left(\frac{1}{1+x^4} - \frac{1}{2} \right) .dx$$

$$= -c \int_{-1}^{+1} \left(\frac{e^{\beta x}}{1+x^4} - \frac{e^{\beta x}}{2} \right) .dx$$

$$= -c \int_{-1}^{+1} \left(\frac{e^{\beta x}}{1+x^4} \right) .dx - -c \int_{-1}^{+1} \left(\frac{e^{\beta x}}{2} \right) .dx$$

Equation C31

The singularity manifests when we integrate the subtrahend:-

$$-c \int_{-1}^{+1} \frac{e^{\beta x}}{2} .dx = -c \left[\frac{e^{\beta x}}{2\beta} \right]_{-1}^{+1}$$

Singular Expression C1

Any attempt to shift the origin of the Skew Factor will of course only complicate matters by introducing problems of its own.

When I offered up the minuend of the integrand of Equation C30 to the Wolfram Mathematica® integration engine[C3] the program delivered the result:-

$$\int_{-1}^{+1} \frac{-ce^{\beta x}}{1+x^4} .dx = -\frac{1}{4} c \, \text{Rootsum} \left[\#1^4 + 1 \& , \frac{e^{\#1.\beta} . Ei(\beta x - \beta \#1)}{\#1^3} \& \right]$$

Equation C32

Ei(z) is the Exponential Integral of its argument z defined by:-

$$Ei(z) = \gamma + \ln(z) + \sum_{i=1}^{n} \frac{z^i}{n \times n!}$$

Equation C33

where γ is the Euler-Mascheroni Constant. ($\gamma \approx 0.577215664901533$).

The integrals that I managed to compute using the Equation C32 expression were both complex and numerically-incorrect. Quite possibly I mis-applied the Mathematica® result, but I was satisfied that any method that appealed to exotic functions would be prohibitively expensive for use in environmental geoscience where of course the quality of the data seldom justifies extreme sophistication.

<u>Gaussian-type Quadrature Applications</u>

Gaussian-type quadrature techniques are schemes of numerical integration that seek to generate high-accuracy numerical results using *unevenly* spaced abscissas: That is panels of unequal width. These abscissas are not of course random but judiciously-generated using a mathematical process designed to give exact integrations for certain families of integrand algebras.

In contrast, Simpson's Rule and other Closed-Form Newton-Cotes schemes use *evenly* spaced abscissas that fit polynomials between data points whether the said polynomials are appropriate or not.

Gauss-Legendre Quadrature (GLQ)[C4]

Gauss-Legendre Quadrature, sometimes simply called Gaussian Quadrature, is the original of these techniques and is exact for the family of integrand objects known as Legendre Polynomials. Gaussian Quadrature is about right for most integrands featuring algebraic polynomials and I found that Gauss-Legendre Quadrature on thirteen points was roughly good for Stream Section Area.

GLQ applications are predicated upon knowledge of the Weight, w_0, and also the i = 0...n Abscissas, x_i, and Weights, w_i. These statistics are computed using specialised functions and extensively-published[C4].

The relevant thirteen $\{w_0, (x_i, w_i)\}$ for my trials are tabulated below:-

i	Abscissor, x_i	Weight, w_i
0		0.232551553230874
1	0.230458315955135	0.226283180262897
2	0.448492751036447	0.207816047536889
3	0.642349339440340	0.178145980761946
4	0.801578090733310	0.138873510219787
5	0.917598399222978	0.092121499837728
6	0.984183054718588	0.040484004765316

Table C4
Gauss-Legendre Abscissas and Weights
for n = 13

If we define the (positivised) integrand as:-

$$f(x) = e^{\beta x}\left(\frac{1}{1+x^4} - \frac{1}{2}\right)$$

Equation C34

and the predicates:-

$$n = 13 \qquad\qquad m = \mathrm{entier}\!\left(\frac{n}{2}\right)$$

Equation C35a   **Equation C35b**

then the appropriate Gauss-Legendre Integral is:-

$$A_{GL} = c\left[\sum_{i=1}^{m} w_i . f(-x_i) + \sum_{i=1}^{m} w_i . f(x_i) + \frac{w_0}{2}\right]$$

Equation C36

As with the Simpson versus Trapezoidal trial above we will document Gaussian-type trials for $c = +0.67$ and $\beta = +1.37$.

For $n = 13$ and these parameters, A_{GL} was 0.59539922178237. The MATHCAD® intrinsic integration, A, of Equation C30 gave 0.595399515241958. Accordingly, the Specific Defect for $(A_{GL}-A)/A$ was -0.000049287844661%.

For a variety of extreme and moderate c and β values specific defect was found to remain less than plus or minus one percent *for the case of areal integration.*

The process was repeated for the curve length line integral (wetted perimeter) defined by the differential of the Equation C30 integrand (with constant c incorporated):-

$$f(x) = \sqrt{1 + \left[ce^{\beta x}\left[\beta\left(\frac{1}{1+x^4} - \frac{1}{2}\right) - \frac{4x^3}{\left(1+x^4\right)^2}\right]\right]^2}$$

Equation C37

For this case, the MATHCAD® intrinsic was 2.451086619650186 and the GLQ approximation 2.31152099922468, yielding a specific defect of -5.694030529423933%.

The wetted perimeter GLQ defect was found to lie between $\pm1\%$ and $\pm10\%$ for a variety of c and β: Values unacceptable to me, so the hunt was on for a cheap and accurate method for both A_w and P_w.

Chebyshev-Gauss Quadrature

Chebyshev-Gauss Quadrature[C5] (CGQ) is optimised for a family of integrands involving the reciprocal square-root of an algebraic polynomial.

An attraction of Chebyshev-Gauss Quadrature is that its Abscissas and Weights can be computed using very simple functions.

The Number of Abscissas of Integration, n, can be defined arbitrarily. My ancient MATHCAD® Student Edition package limits me to an array of fifty elements for $i = 0...49$. Accordingly, I defined $n = 48$ and $k = 1...n$.

All of the Weights, w_k, are fixed as simply:-

$$w = w_0 = \frac{\pi}{n+1}$$

Equation C38

whilst the Abscissas, ξ_k, are given by:-

$$\xi_k = \cos\left[\frac{\left(k - \frac{1}{2}\right).\pi}{n+1}\right]$$

Equation C39

In addition:-

$$g(\xi_k) = \sqrt{1 - \xi_k^2}.f(\xi_k)$$

Equation C40

Taking f(x) as the integrand Equation C34 we may write the Chebyshev-Gauss Integral, A_{CG}, as:-

$$A_{CG} = c\sum_{k=1}^{n} w_k.g(\xi_k) = cw\sum_{k=1}^{n} g(\xi_k) = \frac{c.\pi}{n+1}.\sum_{k=1}^{n}\left[\sqrt{1 - \xi_k^2}.f(\xi_k)\right]$$

Equation C41

In its fully-expanded form, A_{CG}, may be written as:-

$$A_{CG} = \frac{c.\pi}{n+1}.\sum_{k=1}^{n}\left[\sqrt{1 - \cos\left[\frac{\left(k - \frac{1}{2}\right).\pi}{n+1}\right]^2}.f\left(\cos\left[\frac{\left(k - \frac{1}{2}\right).\pi}{n+1}\right]\right)\right]$$

Equation C42

Using c = 0.67, β = +1.37 as before, A_{CG} was found to be 0.595398874290055 yielding the slightly-vitiated defect of −0.000107650726432%.

By applying Chebyshev-Gauss Quadrature to the wetted perimeter integrand Equation C37 (with constant c removed) I determined a value for that curve of 2.449660031101819. This leads to a specific defect on the intrinsic S of −0.058202290238578%, which is an excellent figure for a geomorphological result.

Using these CGQ outcomes, R_h, the Hydraulic Radius, was calculated to be 0.243053675502169, with a specific defect of $+0.058128471613913\%$.

There is currently insufficient evidence to claim that CGQ is superior to 48-panel Simpson's One Third Rule for these parameters.

For a MATHCAD® intrinsic Area, A, and a Simpson's A_{simp} for $n = 48$, the Specific Defect $(A-A_{simp})/A_{simp}$ proved to be -0.000107650726432. This is worse than the *area* GLQ result but well within the limits of metrical error.

A Note on Lobatto Quadrature

For a small number of trial c and β values Gauss-Lobatto Quadrature proved highly accurate for *sectional areas* but unusable for *wetted perimeters*.

<u>References</u>

C1   "CRC Standard Curves and Surfaces"
David Henry von Seggern
CRC Press Incorporated of Boca Raton   1993
ISBN 0-8493-0196-3
pp388

     (Algebraic Curves) Function with a^4+x^4
     2.7.1
     Page 54
     $y = c/(a^4+x^4)$

C2   "Teach Yourself Calculus"
P Abbott BA
The English Universities Press of London EC1
1940   (corrected edition 1963)
pp 380

     Curve Length Integral on Page 275

C3   WolframAlpha
Integral Calculator: Wolfram Mathematica Online Integrator
http://integrals.wolfram.com

C4   Gaussian Quadrature Weights and Abscissae
http://pomax.github.io/bezierinfo/legendre-gauss.html

     (All Gauss-Legendre coefficients for $n < 65$)

C5   Rice University Lecture
CAAM 453/553 – NUMERICAL ANALYSIS I
Lecture 26: More on Gaussian Quadrature
http://www.caam.rice.edu/~embree/caam453/lecture26.pdf

CHAPTER THREE

The Thalean Triangle

Some Properties of the Thalean Triangle

by
James R Warren BSc MSc PhD PGCE

A thalean triangle is a right-angled triangle whose hypotenuse is a circle's diameter. The right angle is opposite that hypotenuse, and is conserved whatever the values of the other angles.

The relevant geometrical relations and notation are illustrated in Figure One a:-

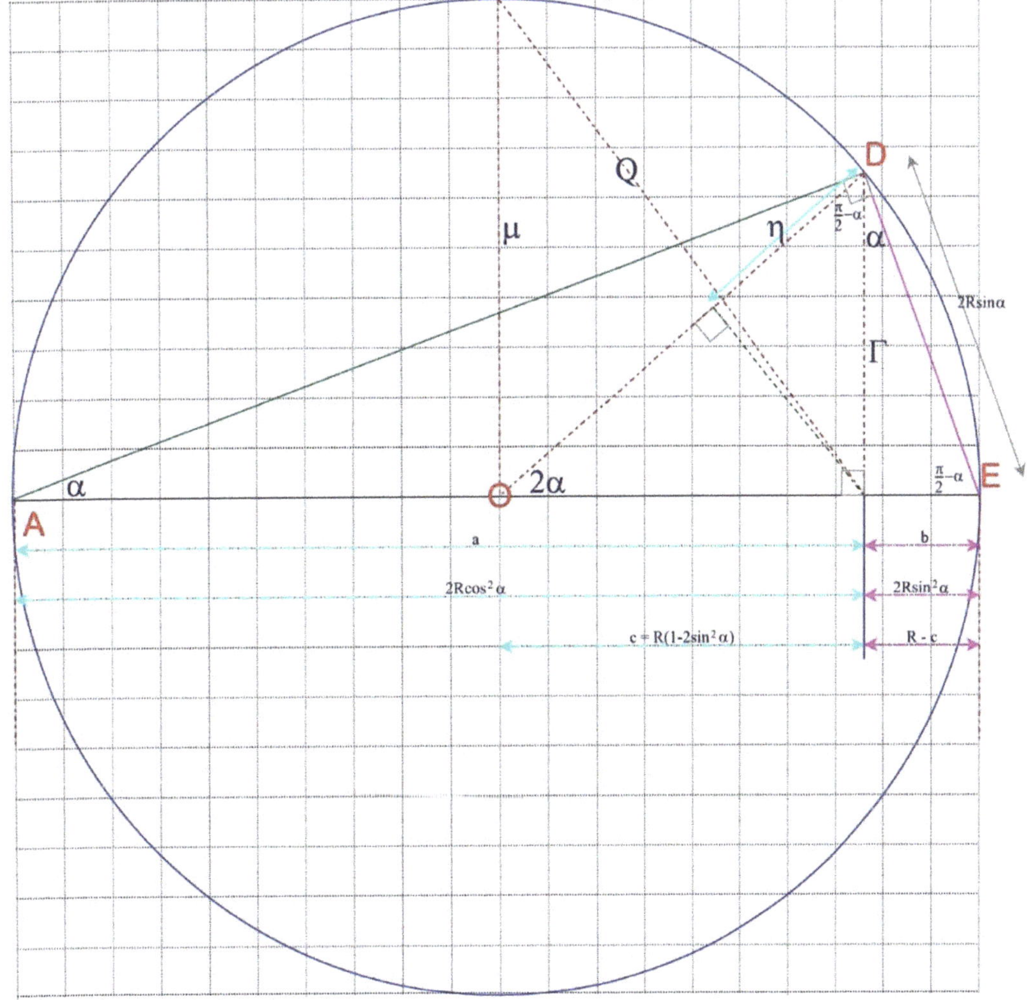

Figure 1a
The Analytic Geometry of the Thalean Triangle

The properties of the thalean triangle may be described in terms of the Radius of the Circumscriptive Circle, R, and the Angle of Elevation, α, only.

By the Sine Rule the Thalean Adjacent, AD, and the Thalean Opposite, DE, are respectively given by:-

$$AD = 2R\cos\alpha$$
Equation 1.1

and:-

$$DE = 2R\sin\alpha$$
Equation 1.2

Furthermore, the Major Diametric, a, is given by:-

$$a = 2R - b = 2R\cos^2(\alpha)$$
Equation 1.3

and:-

$$b = 2R\sin^2(\alpha)$$
Equation 1.4

Equations 1.3 and 1.4 may be restated as:-

$$\cos(\alpha) = \frac{a}{AD}$$
Equation 1.5

and:-

$$\sin(\alpha) = \frac{b}{DE}$$
Equations 1.6

Also, the part-radius lineament c may be defined as:-

$$c = R - b = R - 2R\sin^2(\alpha) = R(1 - 2\sin(\alpha)^2)$$
Equation 1.7

<u>Descriptive Means</u>

Certain Pythagorean means involving the Major and Minor Diametrics may be specified and these are depicted in Figure One.

The Arithmetic Mean, μ

The Arithmetic Mean, μ, is specified in terms of the above equations by:-

$$\mu = \frac{a+b}{2} = \frac{2R\cos^2(\alpha) + 2R\sin^2(\alpha)}{2} = R(\cos^2(\alpha) + \sin^2(\alpha))$$

Equation 1.8

The Harmonic Mean, η

The Harmonic Mean, η, may be expressed thus:-

$$\eta = \frac{2ab}{a+b} = \frac{2ab}{2R} = \frac{2 * 2R\cos^2(\alpha) * 2R\sin^2(\alpha)}{2 * R}$$

Equation 1.9

from which it is apparent that:-

$$\eta = 4R\cos^2(\alpha)\sin^2(\alpha)$$

Equation 1.10

The Geometric Mean, Γ

The Geometric Mean, Γ, is given by:-

$$\Gamma = \sqrt{2R\cos^2(\alpha)\sin^2(\alpha)} = \sqrt{4R^2\cos^2(\alpha)\sin^2(\alpha)}$$

Equation 1.11

which simplifies to:-

$$\Gamma = 2R\cos^2(\alpha)\sin^2(\alpha)$$

Equation 1.12

which is the same as:-

$$\Gamma = R\sin^2(2\alpha)$$

Equation 1.13

The Harmonic, Geometric and Arithmetic Means relate according to:-

$$\eta = \frac{\Gamma^2}{\mu}$$
Equation 1.14

Appropriate substitutions yield:-

$$\eta = \frac{4R^2 \cos^2(\alpha) \sin^2(\alpha)}{R(\cos^2(\alpha) + \sin^2(\alpha))}$$
Equation 1.15

Cancelling the redundant R gives:-

$$\eta = \frac{4R \cos^2(\alpha) \sin^2(\alpha)}{\cos^2(\alpha) + \sin^2(\alpha)}$$
Equation 1.16

The Quadratic Mean, Q

The Quadratic Mean, Q, may be specified as:-

$$Q = \sqrt{\frac{a^2 + b^2}{2}}$$
Equation 1.17

Substitution for a and b then gives-

$$Q = \sqrt{\frac{(2R \cos^2(\alpha))^2 + (2R \sin^2(\alpha))^2}{2}}$$

$$Q = \sqrt{\frac{(4R^2 \cos^4(\alpha)) + (4R^2 \sin^4(\alpha))}{2}}$$

$$Q = \sqrt{\frac{4R^2(\cos^4(\alpha) + \sin^4(\alpha))}{2}}$$

$$Q = \sqrt{2R^2[\cos^4(\alpha) + \sin^4(\alpha)]}$$
Equations 1.18

or:-

$$Q = \sqrt{2}.\,R.\,\sqrt{\cos^4(\alpha) + \sin^4(\alpha)}$$
Equation 1.19

Noting that:-

$$\cos^4(\alpha) = \frac{1}{8}.\,(3 + 4\cos(2\alpha) + \cos(4\alpha))$$
Equation 1.20

$$\sin^4(\alpha) = \frac{1}{8}.\,(3 - 4\cos(2\alpha) + \cos(4\alpha))$$
Equation 1.21

We may recast Q by reference to Equation 1.19 as:-

$$Q = \sqrt{2}.\,R.\,\sqrt{\frac{1}{8}.\,(3 + 4\cos(2\alpha) + \cos(4\alpha)) + \frac{1}{8}.\,(3 - 4\cos(2\alpha) + \cos(4\alpha))}$$
Equation 1.22

This expansion condenses to:-

$$Q = R.\,\sqrt{2}.\,\sqrt{\frac{6}{8} + \frac{2}{8}.\cos(4\alpha)}$$
Equation 1.23

$$Q = R.\,\sqrt{2}.\,\sqrt{\frac{3}{4} + \frac{\cos(4\alpha)}{4}}$$
Equation 1.24

$$Q = \frac{R.\,\sqrt{2}}{2}.\,\sqrt{3 + \cos(4\alpha)}$$
Equation 1.25

$$Q = \frac{R}{\sqrt{2}}.\,\sqrt{3 + \cos(4\alpha)}$$
Equation 1.26

<u>The Area of the Thalean Triangle</u>

Firstly, define the Side Product, s, as:-

$$s = AD.DE.AE$$
Equation 1.27

Then the Area of the Thalean Triangle ADE is given by:-

$$A_{ADE} = \frac{s}{4R}$$
Equation 1.28

By substitution:-

$$A_{ADE} = \frac{AD.DE.AE}{4R}$$
Equation 1.29

$$A_{ADE} = \frac{2.R.\cos(\alpha).2.R.\sin(\alpha).2.R}{4R}$$
Equation 1.30

$$A_{ADE} = \frac{2.R.\cos(\alpha).2.R.\sin(\alpha)}{2}$$
Equation 1.31

$$A_{ADE} = 2R^2 \cos(\alpha) \sin(\alpha)$$
Equation 1.32

Also:-

$$A_{ADE} = R\Gamma$$
Equation 1.33

$$A_{ADE} = R^2 \sin(2\alpha)$$
Equation 1.34

Partitions of the Thalean Triangle:
Sections from the Center

by
James R Warren BSc MSc PhD PGCE

This disquisition discusses the partition of a thalean triangle into two component triangles and a quadrilateral by two lines radiating from the circle center.

The relevant geometrical relations and notation are illustrated in Figure One b:-

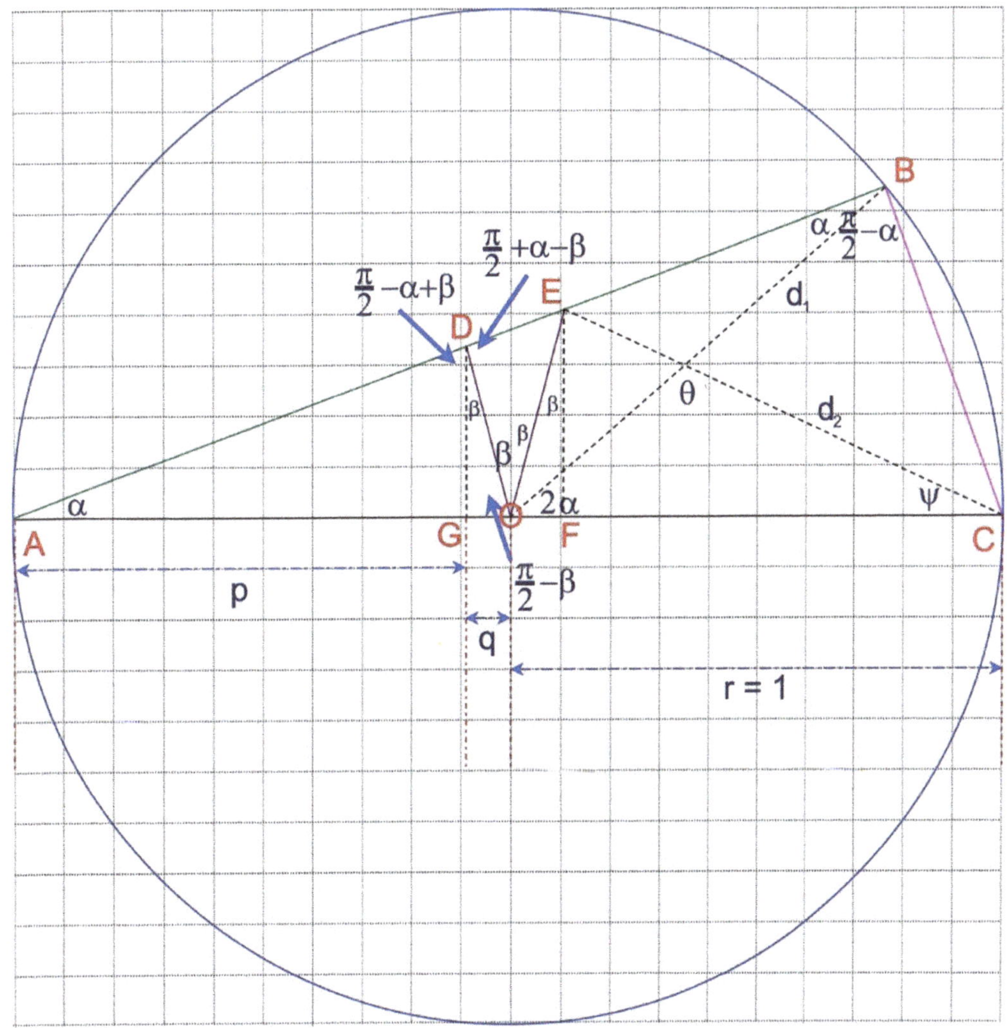

Figure 1b
The Analytic Geometry of the Thalean Triangle

The properties of the partitioning system may be described in terms of the Radius of the Circumscriptive Circle, R, and the Angle of Elevation, α, together with the Angle of Section, β. Though Figure One shows the radius in terms of unity as r = 1, the method is general enough to embrace the general radius R.

<u>Some Line Definitions</u>

By the Sine Rule:-

$$DO = \frac{R \sin{(\alpha)}}{\cos{(\alpha - \beta)}}$$

Equation 3.1

whilst:-

$$DG = DO.\cos(\beta) = \frac{R \sin(\alpha).\cos{(\beta)}}{\cos{(\alpha - \beta)}}$$

Equation 3.2

Accordingly:-

$$p = \frac{DG}{\tan(\alpha)} = \frac{R \sin(\alpha).\cos{(\beta)}}{\tan(\alpha).\cos{(\alpha - \beta)}} = \frac{R \cos(\alpha)\cos{(\beta)}}{\cos{(\alpha - \beta)}}$$

Equation 3.3

and:-

$$q = R - p = R\left(1 - \frac{\cos(\alpha)\cos(\beta)}{\cos(\alpha - \beta)}\right) = DG.\tan{(\beta)}$$

Equation 3.4

It therefore follows that:-

$$q = DG.\tan(\beta) = \frac{R.\sin(\alpha).\cos(\beta).\tan{(\beta)}}{\cos(\alpha - \beta)}$$

Equation 3.5

or:-

$$q = \frac{R.\sin(\alpha).\sin(\beta)}{\cos(\alpha - \beta)}$$

Equation 3.6

$$DE = DO.\left(\frac{\sin(2\beta)}{\cos(\alpha+\beta)}\right) = \frac{R.\sin(\alpha)}{\cos(\alpha-\beta)} \cdot \frac{\sin(2\beta)}{\cos(\alpha+\beta)}$$
Equation 3.7

or:-

$$DE = \frac{R.\sin(\alpha)\sin(2\beta)}{\cos^2(\alpha) + \cos^2(\beta) - 1}$$
Equation 3.8

We may further define:-
$$AO = OC = R$$
Equation 3.9

$$AC = 2R$$
Equation 3.10

$$AB = 2R\cos(\alpha)$$
Equation 3.11

$$BC = 2R\sin(\alpha)$$
Equation 3.12

$$AD = \frac{R.\cos(\beta)}{\cos(\alpha-\beta)}$$
Equation 3.13

From the geometry of Figure One:-

$$EB = AB - AD - DE$$
Equation 3.12

From which a possible assembly is:-

$$EB = 2R\cos(\alpha) - \frac{R.\sin(\alpha)}{\cos(\beta-\alpha)} \cdot \frac{\cos(\beta)}{\sin(\alpha)} - \frac{R.\sin(\alpha)}{\cos(\beta-\alpha)} \cdot \frac{\sin(2\beta)}{\cos(\alpha+\beta)}$$
Equation 3.13

Manipulation then gives the useful simplification:-

$$EB = R \cdot \left[2\cos(\alpha) - \left(\frac{\cos(\beta)}{\cos(\beta - \alpha)} \right) \cdot \left(1 + \frac{2\sin(\alpha)\sin(\beta)}{\cos(\alpha + \beta)} \right) \right]$$

Equation 3.14

The line AE may be specified as:-

$$AE = AD + DE = \frac{R.\cos(\beta)}{\cos(\alpha - \beta)} + \frac{R.\sin(\alpha)}{\cos(\alpha - \beta)} \cdot \frac{\sin(2\beta)}{\cos(\alpha + \beta)}$$

Equation 3.15

or:-

$$AE = \frac{R}{\cos(\alpha - \beta)} \cdot \left(\cos(\beta) + \frac{\sin(\alpha)\sin(2\beta)}{\cos(\alpha + \beta)} \right)$$

Equation 3.16

Now by the Sine Rule:-

$$EO = AE \cdot \frac{\sin(\alpha)}{\cos(-\beta)}$$

Equation 3.17

and note that:-

$$\cos(-\beta) = \cos(\beta)$$

Equation 3.18

Therefore:-

$$EO = \frac{R.\sin(\alpha)}{\cos(\beta)\cos(\alpha - \beta)} \cdot \left(\cos(\beta) + \frac{\sin(\alpha)\sin(2\beta)}{\cos(\alpha + \beta)} \right) \cdot$$

Equation 3.19

Which reduces to:-

$$EO = \frac{R.\sin(\alpha)}{\cos(\alpha + \beta)}$$

Equation 3.20

Also:-

$$EF = EO.\cos(\beta) = \frac{R.\sin(\alpha).\cos(\beta)}{\cos(\alpha + \beta)}$$

Equation 3.21

<u>Component Areas</u>

Triangle ADO

$$ADO = \frac{1}{2} \cdot AO \cdot AD \cdot \sin(\alpha)$$

Equation 3.22

From which by substitution we may develop:-

$$ADO = \frac{1}{2} \cdot R \cdot \frac{R.\cos(\beta)}{\cos(\alpha - \beta)} \cdot \sin(\alpha)$$

Equation 3.23

$$ADO = \frac{R^2.\sin(\alpha).\cos(\beta)}{2.\cos(\alpha - \beta)}$$

Equation 3.24

Furthermore:-

$$ADO = \frac{R}{2}DG = \frac{R}{2}\frac{R\sin(\alpha).\cos(\beta)}{\cos(\alpha - \beta)} \equiv Eqn.\,3.23$$

Equation 3.25

Triangle ODE

$$ODE = \frac{1}{2} \cdot DO \cdot EO \cdot \sin(2\beta)$$

Equation 3.26

From which by substitution we may develop:-

$$ODE = \frac{1}{2} \cdot \frac{R\sin(\alpha)}{\cos(\alpha - \beta)} \cdot \frac{R.\sin(\alpha)}{\cos(\alpha + \beta)} \cdot \sin(2\beta)$$

Equation 3.27

or:-

$$ODE = \frac{R^2}{2} \cdot \frac{\sin^2(\alpha) \cdot \sin(2\beta)}{\cos(\alpha + \beta) \cdot \cos(\alpha - \beta)}$$

Equation 3.28

Noting that:-

$$2\cos(\alpha + \beta) \cdot \cos(\alpha - \beta) = \cos(2\alpha) + \cos(2\beta)$$

Equation 3.29

$$ODE = \frac{R^2 \sin^2(\alpha) \cdot \sin(2\beta)}{\cos(2\alpha) + \cos(2\beta)}$$

Equation 3.30

Quadrilateral OEBC

One way to compute the Area of Quadrilateral OEBC is by the addition of component right-angled and scalene triangles:-

$$OEBC = \frac{EF \cdot OC}{2} + \frac{EB \cdot BC}{2} = \frac{1}{2} \cdot (R.EF + EB.BC)$$

Equation 3.31

Appropriate substitutions give:-

$$OEBC = \frac{1}{2} \cdot \left(R. \frac{R.\sin(\alpha) \cdot \cos(\beta)}{\cos(\alpha + \beta)} \right.$$
$$\left. + R \left[2\cos(\alpha) - \left(\frac{\cos(\beta)}{\cos(\beta - \alpha)} \right) \cdot \left(1 + \frac{2\sin(\alpha)\sin(\beta)}{\cos(\alpha + \beta)} \right) \right] . 2R\sin(\alpha) \right)$$

Equation 3.32

Manipulation allows us to write:-

$$OEBC = \frac{R^2 \sin(\alpha)}{2} \cdot \left[\frac{\cos(\beta)}{\cos(\alpha + \beta)} + 2 \left\{ 2\cos(\alpha) - \left(\frac{\cos(\beta)}{\cos(\beta - \alpha)} \right) \cdot \left(1 + \frac{2\sin(\alpha)\sin(\beta)}{\cos(\alpha + \beta)} \right) \right\} \right]$$

Equation 3.33

Further simplifications allow us to quote the Area OEBC in the form:-

$$OEBC = \frac{R^2 \sin(\alpha) \cos(\beta)}{2\cos(\alpha + \beta)} \cdot \left(4\cos(\alpha + \beta) \cdot \frac{\cos(\alpha)}{\cos(\beta)} - 1 \right)$$

Equation 3.34

The Total Area of the Thalean Triangle ABC

The Total Area of the Thalean Triangle may be specified in many, perhaps infinite, ways.

Let A_T be The Area of the Thalean Triangle then the alternatives interesting to us include:-

$$A_T = ADO + ODE + OEBC = \frac{AB \cdot BC}{2} = \frac{\sin(\alpha) \cdot AB \cdot AC}{2}$$

Equation 3.35

Of these alternatives let us focus upon the first:-

$$A_T = ADO + ODE + OEBC$$

Equation 3.36

And effect the following substitutions:-

$$A_T = \frac{R^2 . \sin(\alpha) . \cos(\beta)}{2 . \cos(\alpha - \beta)} + \frac{R^2}{2} \cdot \frac{\sin^2(\alpha) . \sin(2\beta)}{\cos(\alpha + \beta) . \cos(\alpha - \beta)} + \frac{R^2 \sin(\alpha) \cos(\beta)}{2\cos(\alpha + \beta)}$$
$$\cdot \left(4\cos(\alpha + \beta) \cdot \frac{\cos(\alpha)}{\cos(\beta)} - 1 \right)$$

Equation 3.37

Suitable segregations permit the simplification:-

$$A_T = \frac{R^2}{2} \cdot \sin(\alpha)$$
$$\cdot \left[4\cos(\alpha) + \cos(\beta) \cdot \left\{ \frac{1}{\cos(\alpha - \beta)} - \frac{1}{\cos(\alpha + \beta)} \right\} + \frac{\sin(\alpha)\sin(2\beta)}{\cos(\alpha + \beta) \cdot \cos(\alpha - \beta)} \right]$$

Equation 3.38

Noting the identity:-

$$\cos(\beta) \cdot \left\{ \frac{1}{\cos(\alpha - \beta)} - \frac{1}{\cos(\alpha + \beta)} \right\} = \frac{-\sin(\alpha).\sin(2\beta)}{\cos(\alpha + \beta).\cos(\alpha - \beta)}$$

Equation 3.39

We may simplify Equation 3.38 to:-

$$A_T = \frac{R^2}{2} \cdot \sin(\alpha) \cdot 4\cos(\alpha) = 2R^2 \sin(\alpha) \cos(\alpha)$$

Equation 3.40

whose latter form is the equation for Thalean Triangle Area developed in my late paper THALPROP "Some Properties of the Thalean Triangle". A proof by tautology. It is nice to check the integrity of things so far.

Several other combinations are mathematically viable but I remain unconvinced of their superior computational economy.

Fractional Areas

There is an interest in treating the three component figures of the thalean triangle as fractions of unity.

Let us note the respective Fractional Areas of ADO, ODE and OEBC respectively as FADO, FODE and FOEBC.

Then:-

$$1 = FADO + FODE + FOEBC$$

Equation 3.41

that shall develop a grand identity.

We will for convenience employ the THALPROP form of Thalean Triangle Area:-

$$A_T = 2R^2 \cos(\alpha) \sin(\alpha)$$

Equation 3.42

(The order of terms is of course immaterial).

Accordingly:-

$$FADO = \frac{ADO}{A_T} = \frac{1}{A_T} \cdot ADO = \frac{1}{2R^2 \cos(\alpha) \sin(\alpha)} \cdot \frac{R^2.\sin(\alpha).\cos(\beta)}{2.\cos(\alpha - \beta)}$$

Equation 3.43

$$FODE = \frac{ODE}{A_T} = \frac{1}{A_T} \cdot ODE = \frac{1}{2R^2 \cos(\alpha)\sin(\alpha)} \cdot \frac{R^2}{2} \cdot \frac{\sin^2(\alpha).\sin(2\beta)}{\cos(\alpha+\beta).\cos(\alpha-\beta)}$$

Equation 3.44

$$FOEBC = \frac{OEBC}{A_T} = \frac{1}{A_T} \cdot OEBC$$

$$= \frac{1}{2R^2 \cos(\alpha)\sin(\alpha)} \cdot \frac{R^2 \sin(\alpha)\cos(\beta)}{2\cos(\alpha+\beta)} \cdot \left(4\cos(\alpha+\beta) \cdot \frac{\cos(\alpha)}{\cos(\beta)} - 1 \right)$$

Equation 3.45

Appropriate simplifications of Equations 3.43, 3.44 and 3.45 allow us to write:-

$$FADO = \frac{1}{4} \cdot \frac{\cos(\beta)}{\cos(\alpha) \cdot \cos(\alpha-\beta)}$$

Equation 3.46

$$FODE = \frac{1}{4} \cdot \frac{\sin(\alpha) \cdot \cos(2\beta)}{\cos(\alpha) \cdot \cos(\alpha+\beta) \cdot \cos(\alpha-\beta)}$$

Equation 3.47

$$FOEBC = 1 - \frac{1}{4} \cdot \frac{\cos(\beta)}{\cos(\alpha) \cdot \cos(\alpha+\beta)}$$

Equation 3.48

Therefore, by Equation 3.41:-

$$1 = \frac{1}{4} \cdot \frac{\cos(\beta)}{\cos(\alpha) \cdot \cos(\alpha-\beta)} + \frac{1}{4} \cdot \frac{\sin(\alpha) \cdot \cos(2\beta)}{\cos(\alpha) \cdot \cos(\alpha+\beta) \cdot \cos(\alpha-\beta)} + 1 - \frac{1}{4}$$
$$\cdot \frac{\cos(\beta)}{\cos(\alpha) \cdot \cos(\alpha+\beta)}$$

Equation 3.49

This may be expressed as:-

$$0 = \frac{1}{4} \cdot \frac{\cos(\beta)}{\cos(\alpha) \cdot \cos(\alpha-\beta)} + \frac{1}{4} \cdot \frac{\sin(\alpha) \cdot \cos(2\beta)}{\cos(\alpha) \cdot \cos(\alpha+\beta) \cdot \cos(\alpha-\beta)} - \frac{1}{4}$$
$$\cdot \frac{\cos(\beta)}{\cos(\alpha) \cdot \cos(\alpha+\beta)}$$

Equation 3.50

which simplifies to the interesting crossover identity:-

$$0 = \frac{\cos(\beta)}{\cos(\alpha - \beta)} + \frac{\sin(\alpha) \cdot \cos(2\beta)}{\cos(\alpha + \beta) \cdot \cos(\alpha - \beta)} - \frac{\cos(\beta)}{\cos(\alpha + \beta)}$$

Equation 3.51

Some Convenient Identities

by
James R Warren BSc MSc PhD PGCE

There follow some trigonometrical identities which are, or may be, of use in the simplification or economisation of thalean triangle partition solution structures.

These identities are at least applicable to angles between zero and π/4 radians.

(a)

$$\frac{1 + \tan(\gamma)}{\tan(\gamma) - 1} \equiv \frac{1}{\tan\left(\gamma - \frac{\pi}{4}\right)}$$

(b)

$$\cos(\gamma) - \sin(\gamma) \equiv \frac{2\sin\left(\frac{\pi}{4} - \gamma\right)}{\sqrt{2}} \equiv \sqrt{2} \cdot \sin\left(\frac{\pi}{4} - \gamma\right) \equiv \cos(\gamma)(1 - \tan(\gamma))$$

(c)

$$\cos(\gamma) + \sin(\gamma) \equiv \frac{2\sin\left(\frac{3\pi}{4} - \gamma\right)}{\sqrt{2}} \equiv \sqrt{2} \cdot \sin\left(\frac{3\pi}{4} - \gamma\right) \equiv \cos(\gamma)(1 + \tan(\gamma))$$

(d)

$$\sin\left(\frac{\pi}{4} + \gamma\right) \equiv \frac{1}{\sqrt{2}} \cdot \cos(\gamma) \cdot (1 + \tan(\gamma))$$

(e)

$$1 + \tan(\gamma) = \frac{\cos(\gamma) + \sin(\gamma)}{\cos(\gamma)}$$

(f)

$$1 - \tan(\gamma) = \frac{\cos(\gamma) - \sin(\gamma)}{\cos(\gamma)}$$

(g)

$$1 - \tan^2(\gamma) = \frac{2\tan(\gamma)}{\tan(2\gamma)} = (1 + \tan(\gamma))(1 - \tan(\gamma)) = \frac{\cos^2(\gamma) - \sin^2(\gamma)}{\cos^2(\gamma)}$$

(h)

$$\tan(\gamma) - 1 = \frac{-\sqrt{2}\cos\left(\frac{\pi}{4} + \gamma\right)}{\cos(\gamma)}$$

(i)

$$\frac{1 + \tan(\gamma)}{1 - \tan(\gamma)} \equiv \frac{\sin\left(\frac{\pi}{4} + \gamma\right)}{\sin\left(\frac{\pi}{4} - \gamma\right)} \equiv \frac{\cos(\gamma) + \sin(\gamma)}{\cos(\gamma) - \sin(\gamma)} \equiv \tan\left(\frac{\pi}{4} + \gamma\right) \equiv \frac{1}{\cos(2\gamma)} + \tan(2\gamma)$$

(j)

$$\frac{1 - \tan(\gamma)}{1 + \tan(\gamma)} \equiv \frac{\sin\left(\frac{\pi}{4} - \gamma\right)}{\sin\left(\frac{\pi}{4} + \gamma\right)} \equiv \frac{\cos(\gamma) - \sin(\gamma)}{\cos(\gamma) + \sin(\gamma)} \equiv \tan\left(\frac{\pi}{4} - \gamma\right) \equiv \frac{1}{\cos(2\gamma)} - \tan(2\gamma)$$

(k)

$$\frac{-\sin(\alpha)\sin(2\beta)}{\cos(\alpha+\beta).\cos(\alpha-\beta)} \equiv \cos(\beta).\left(\frac{1}{\cos(\alpha-\beta)} - \frac{1}{\cos(\alpha+\beta)}\right) \equiv \frac{-2\sin(\alpha)\sin(\beta)\cos(\beta)}{\cos(\alpha+\beta).\cos(\alpha-\beta)}$$

(l)

$$\left(\frac{1}{\cos(\alpha-\beta)} - \frac{1}{\cos(\alpha+\beta)}\right) \equiv \frac{-2\sin(\alpha)\sin(\beta)\cos(\beta)}{\cos(\beta).\cos(\alpha+\beta).\cos(\alpha-\beta)}$$

(m)

$$\left(\frac{1}{\cos(\alpha+\beta)} - \frac{1}{\cos(\alpha-\beta)}\right) \equiv \frac{2\sin(\alpha)\sin(\beta)\cos(\beta)}{\cos(\beta).\cos(\alpha+\beta).\cos(\alpha-\beta)} \equiv \frac{2\sin(\alpha)\sin(\beta)}{\cos(\alpha+\beta).\cos(\alpha-\beta)}$$

(n)

$$2\sin(\alpha)\sin(\beta) \equiv \cos(\alpha+\beta).\cos(\alpha-\beta).\left(\frac{1}{\cos(\alpha+\beta)} - \frac{1}{\cos(\alpha-\beta)}\right)$$

(o)[1]

$$\cos(\alpha+\beta).\cos(\alpha-\beta) \equiv \frac{2\sin(\alpha)\sin(\beta)}{\left(\frac{1}{\cos(\alpha+\beta)} - \frac{1}{\cos(\alpha-\beta)}\right)}$$
$$\equiv \cos^2(\alpha) + \cos^2(\beta) - 1 \equiv \cos^2(\alpha) - \sin^2(\beta)$$

(p)

$$0 \equiv \frac{\cos(\beta)}{\cos(\alpha-\beta)} + \frac{\sin(\alpha).\sin(2\beta)}{\cos(\alpha+\beta).\cos(\alpha-\beta)} - \frac{\cos(\beta)}{\cos(\alpha+\beta)}$$

(q)

$$0 \equiv \frac{\sin(\alpha).\sin(2\beta)}{\cos(\alpha+\beta).\cos(\alpha-\beta)} + \cos(\beta).\left(\frac{1}{\cos(\alpha-\beta)} - \frac{1}{\cos(\alpha+\beta)}\right)$$

(r)

$$0 \equiv \frac{\sin(\gamma)}{2\cos(\alpha)}\left[\frac{1}{\cos(\alpha-\gamma)}\left\{\frac{1}{\sin(2\gamma)} - \tan(\gamma)\right\} + \frac{(\cos(\alpha)+\sin(\alpha))(\cos(\gamma)+\sin(\gamma))}{\cos(\alpha+\gamma)\cos(\alpha-\gamma)}\right.$$
$$\left. - \frac{1}{\cos(\alpha+\gamma)}\left\{1 + \frac{1}{\sin(2\gamma)}\right\}\right]$$

(s)

$$1 \equiv \frac{1}{4}\cdot\frac{\cos(\gamma).(1-\tan(\gamma))^2}{\cos(\alpha).\cos(\alpha-\gamma)} + \frac{2\tan(\alpha)\sin^2(\gamma)}{\tan(2\gamma).[\cos(2\alpha)+\cos(2\gamma)]}$$
$$\cdot\left[1 + \frac{1}{\sin(\alpha).\cos(\gamma)}\cdot\left(\frac{\cos(\alpha+\gamma)}{1+\tan(\gamma)} + \frac{\cos(\alpha-\gamma)}{1-\tan(\gamma)}\right)\right]$$
$$+ \frac{\sin(\gamma)}{\cos(\alpha).\tan(2\gamma)}\cdot\left(\frac{1}{2.\cos(\alpha+\gamma)} + \frac{\cos(\alpha)}{\cos(\gamma)+\sin(\gamma)} - \frac{\sin(\alpha).\tan(\alpha+\gamma)}{\cos(\gamma)-\sin(\gamma)}\right)$$

$(t)^2$

$$\cos(\alpha) \equiv \frac{1}{\sec\alpha} \equiv \frac{1}{\sqrt{1+\tan^2(\alpha)}} \equiv \frac{1}{\sqrt{1+\left(\frac{dy}{dx}\right)^2}}$$

$(u)^2$

$$\sin(\alpha) \equiv \sqrt{1-\cos^2\alpha} \equiv \frac{\frac{dy}{dx}}{\sqrt{1+\left(\frac{dy}{dx}\right)^2}} \equiv \frac{\tan(\alpha)}{\sqrt{1+\tan^2(\alpha)}}$$

$(v)^2$

$$e^{\tanh^{-1}x} \equiv \sqrt{\frac{1+x}{1-x}}$$

(w)

$$e^{\tanh^{-1}\alpha} \equiv \sqrt{\frac{1+\alpha}{1-\alpha}}$$

For $0 \leq \alpha < \pi/4$.

(x)

$$e^{2\tanh^{-1}\alpha} \equiv \frac{1+\alpha}{1-\alpha}$$

For $0 \leq \alpha < \pi/4$.

<u>References</u>

1   "Plane Trigonometry"
For the Use of Colleges and Schools
Eighth Edition   1880
I Todhunter
MacMillan and Company of London
Function Square Identities: p55

2   "Practical Mathematics"
Engineering Degree Series
Louis Toft and ADD McKay
Second Edition   1946
Sir Isaac Pitman and Sons Limited
London
612 pp
(t), (u)   p214 Evolutes
(v)   p57 Limits, etc.

Partitions of the Thalean Triangle:
Sections from the Nadir

by
James R Warren BSc MSc PhD PGCE

This disquisition discusses the partition of a thalean triangle into one component triangle and two quadrilaterals by two lines radiating from the circle nadir.
The relevant geometrical relations and notation are illustrated in Figure One c:-

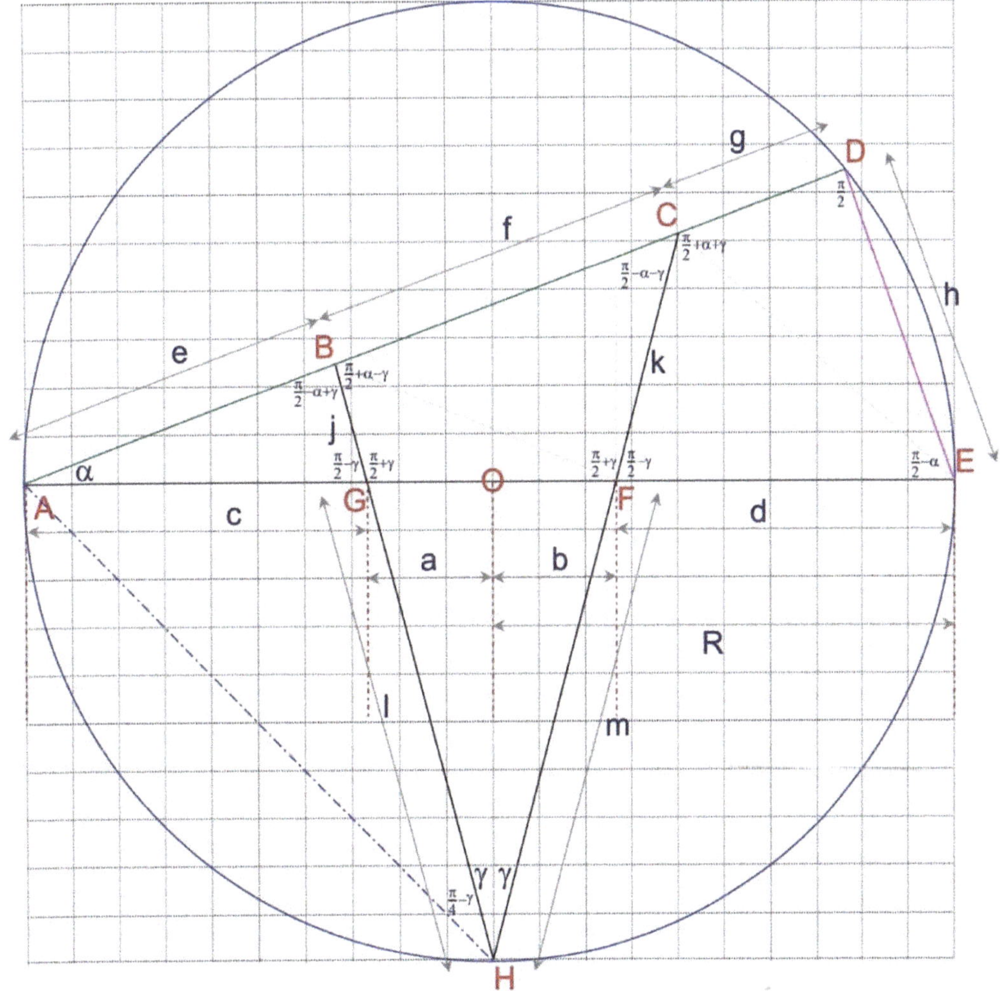

Figure 1c
The Analytic Geometry of the Thalean Triangle

The properties of the partitioning system may be described in terms of the Radius of the Circumscriptive Circle, R, and the Angle of Elevation, α, together with the Angle of Section from the Nadir, γ.

With reference to Figure One:

$$a = b = R \tan(\gamma)$$
Equation 4.1

$$c = d = R(1 - \tan(\gamma))$$
Equation 4.2

$$h = 2R \sin(\alpha)$$
Equation 4.3

$$AE = 2R$$
Equation 4.4

$$AD = 2R \cos(\alpha)$$
Equation 4.5

By the Sine Rule:-

$$e = \frac{\sqrt{2} \cdot R \cdot \sin\left(\frac{\pi}{4} - \gamma\right)}{\cos(\alpha - \gamma)}$$
Equation 4.6

and:-

$$e + f = \frac{\sqrt{2} \cdot R \cdot \sin\left(\frac{\pi}{4} + \gamma\right)}{\cos(\alpha + \gamma)}$$
Equation 4.7

therefore:-

$$f = e + f - e = \frac{\sqrt{2} \cdot R \cdot \sin\left(\frac{\pi}{4} + \gamma\right)}{\cos(\alpha + \gamma)} - \frac{\sqrt{2} \cdot R \cdot \sin\left(\frac{\pi}{4} - \gamma\right)}{\cos(\alpha - \gamma)}$$
Equation 4.8

or:-

$$f = \sqrt{2} \cdot R \cdot \left\{ \frac{\sin\left(\frac{\pi}{4} + \gamma\right)}{\cos(\alpha + \gamma)} - \frac{\sin\left(\frac{\pi}{4} - \gamma\right)}{\cos(\alpha - \gamma)} \right\}$$

Equation 4.9

Furthermore:-

$$g = AD - (e + f) = 2R\cos(\alpha) - \frac{\sqrt{2}R\,\sin\left(\frac{\pi}{4} + \gamma\right)}{\cos(\alpha + \gamma)}$$

Equation 4.10

Simplification yields the form:-

$$g = 2R\left\{ \cos(\alpha) - \frac{1}{2} \cdot \frac{\cos(\gamma) + \sin(\gamma)}{\cos(\alpha + \gamma)} \right\}$$

Equation 4.11

Further applications of the Sine Rule to the conformations of Figure One allow us to specify:-

$$j = R(1 - \tan(\gamma))\frac{\sin(\alpha)}{\cos(\alpha - \gamma)}$$

Equation 4.12

$$k = R(1 + \tan(\gamma))\frac{\sin(\alpha)}{\cos(\alpha + \gamma)}$$

Equation 4.13

from which it may be established that:-

$$\frac{k}{j} = \frac{(1 + \tan(\gamma)) \cdot \cos(\alpha - \gamma)}{(1 - \tan(\gamma)) \cdot \cos(\alpha + \gamma)}$$

Equation 4.14

Also note that:-

$$l = m = \frac{R}{\cos(\gamma)}$$
Equation 4.15

Total and Partition Areas

The Total Area of the Thalean Triangle ADE is, amongst myriad possibilities, given by:-

$$A_{ADE} = \frac{AD.h}{2}$$
Equation 4.16

As noted in previous treatments, a useful alternate form is:-

$$A_{ADE} = 2R^2 \cos(\alpha) \sin(\alpha)$$
Equation 4.17

Triangle AGB

The Area of Triangle AGB is given by:-

$$A_{AGB} = \frac{1}{2}.\sin(\alpha).e.c$$
Equation 4.18

Appropriate substitutions give:-

$$A_{AGB} = \frac{1}{2}.\sin(\alpha).\frac{\sqrt{2}\cdot R\cdot\sin\left(\frac{\pi}{4}-\gamma\right)}{\cos(\alpha-\gamma)}.R(1-\tan(\gamma))$$
Equation 4.19

Simplification of Equation 4.19 yields:-

$$A_{AGB} = \frac{R^2}{2}.\frac{\sin(\alpha)}{\cos(\gamma)}.\frac{[\cos(\gamma)-\sin(\gamma))]^2}{\cos(\alpha-\gamma)}$$
Equation 4.20

Quadrilateral FCBG

The Area of Quadrilateral FCBG may be resolved to those of the two component Triangles FBG and FCB whose respective Areas may conveniently be specified as:

$$A_{FBG} = \frac{1}{2} \cdot \sin\left(\frac{\pi}{2} + \gamma\right) . j . 2 . a$$

Equation 4.21

$$A_{FCB} = \frac{1}{2} \cdot \sin\left(\frac{\pi}{2} - \alpha - \gamma\right) . k . f$$

Equation 4.22

Combination of Equations 4.21 and 4.22 allows us to cite the Quadrilateral Area as:-

$$A_{FCBG} = A_{FBG} + A_{FCB}$$

Equation 4.23

Appropriate substitutions and simplifications enable:-

$$A_{FCBG} = \frac{1}{2} \cdot \cos(-\gamma) . j . 2 . a + \frac{1}{2} \cdot \cos(\alpha + \gamma) . k . f$$

Equation 4.24

Expansion and re-arrangement of this gives us:-

$$A_{FCBG} = \frac{1}{2} R^2 (1 - \tan(\gamma)) \frac{\sin(\alpha)}{\cos(\alpha - \gamma)} \left[2\sin(\gamma) + \sqrt{2}\cos(\alpha - \gamma) \frac{1 + \tan(\gamma)}{1 - \tan(\gamma)} \right.$$
$$\left. \cdot \left\{ \frac{\sin\left(\frac{\pi}{4} + \gamma\right)}{\cos(\alpha + \gamma)} - \frac{\sin\left(\frac{\pi}{4} - \gamma\right)}{\cos(\alpha - \gamma)} \right\} \right]$$

Equation 4.25

or alternatively:-

$$A_{FCBG} = \frac{1}{2} R^2 (1 - \tan(\gamma)) \frac{\sin(\alpha)}{\cos(\alpha - \gamma)} \left[2\sin(\gamma) + \sqrt{2}\cos(\alpha - \gamma) \frac{1 + \tan(\gamma)}{1 - \tan(\gamma)} \right.$$
$$\left. \cdot \left\{ \frac{\cos(\alpha - \gamma)\left(\sin\left(\frac{\pi}{4} + \gamma\right)\right) - \cos(\alpha + \gamma)\left(\sin\left(\frac{\pi}{4} - \gamma\right)\right)}{\cos(\alpha + \gamma) \cdot \cos(\alpha - \gamma)} \right\} \right]$$

Equation 4.26

Now note the identity:-

$$\cos(\alpha + \gamma) \cdot \cos(\alpha - \gamma) = \cos^2(\alpha) - \sin^2(\gamma)$$

Equation 4.27

and also the identity:-

$$\cos(\alpha - \gamma)\left(sin\left(\frac{\pi}{4} + \gamma\right)\right) - \cos(\alpha + \gamma)\left(sin\left(\frac{\pi}{4} - \gamma\right)\right) = \frac{\sqrt{2}}{2} \cdot \sin(2\gamma) \cdot (\cos(\alpha) + \sin(\alpha))$$

Equation 4.28

(which is not of course in its most condensed form).
Then the Quadrilateral Area may, with excision of redundancies, be re-stated as:-

$$A_{FCBG} = R^2(1 - \tan(\gamma))\frac{\sin(\alpha)}{\cos(\alpha - \gamma)}\left[\sin(\gamma) + \cos(\alpha - \gamma)\frac{1 + \tan(\gamma)}{1 - \tan(\gamma)}\right.$$
$$\left. \cdot \left\{\frac{\cos(\gamma)\left(sin(\gamma) \cdot (\cos(\alpha) + \sin(\alpha))\right)}{\cos^2\alpha - \sin^2(\gamma)}\right\}\right]$$

Equation 4.29

Further re-arrangements permit:-

$$A_{FCBG} = R^2\frac{\sin(\alpha)\sin(\gamma)}{\cos(\alpha - \gamma)}\left\{1 - \tan(\gamma) + \cos(\alpha\right.$$
$$\left. - \gamma)\left[\frac{(\cos(\alpha) + \sin(\alpha)) \cdot (\cos(\gamma) + \sin(\gamma))}{(\cos(\alpha) + \sin(\gamma)) \cdot (\cos(\alpha) - \sin(\gamma))}\right]\right\}$$

Equation 4.30

Quadrilateral EDCF

The Area of Quadrilateral ECDF may be resolved to those of the two component Triangles ECF and EDC whose respective Areas may conveniently be specified as:

$$A_{ECF} = \frac{1}{2} \cdot \sin\left(\frac{\pi}{2} - \gamma\right) \cdot d \cdot k$$

Equation 4.31

$$A_{EDC} = \frac{1}{2} \cdot \sin\left(\frac{\pi}{2}\right) \cdot h \cdot g$$
Equation 4.32

Combination of Equations 4.31 and 4.32 allows us to cite the Quadrilateral Area as:-

$$A_{EDCF} = A_{ECF} + A_{EDC}$$
Equation 4.33

Appropriate substitutions and simplifications enable:-

$$A_{EDCF} = \frac{1}{2} \cdot \sin\left(\frac{\pi}{2} - \gamma\right) \cdot d \cdot k - \frac{1}{2} \cdot \sin\left(\frac{\pi}{2}\right) \cdot h \cdot g$$
Equation 4.34

which simplifies to:-

$$A_{EDCF} = \frac{1}{2}[\cos(\gamma)dk + hg]$$
Equation 4.35

appropriate substitutions give:-

$$A_{EDCF} = \frac{1}{2}\left[\cos(\gamma)R(1 - \tan(\gamma))R(1 + \tan(\gamma)) \cdot \frac{\sin(\alpha)}{\cos(\alpha + \gamma)}\right.$$
$$\left. + 2R\sin(\alpha)2R\left\{\cos(\alpha) - \frac{1}{2} \cdot \frac{\cos(\gamma) + \sin(\gamma)}{\cos(\alpha + \gamma)}\right\}\right]$$
Equation 4.36

Further simplifications then yield:-

$$A_{EDCF} = R^2\sin(\alpha)\left[\frac{\sin(\alpha)}{\tan(2\gamma) \cdot \cos(\alpha + \gamma)} + 2\left\{\cos(\alpha) - \frac{1}{2} \cdot \frac{\cos(\gamma) + \sin(\gamma)}{\cos(\alpha + \gamma)}\right\}\right]$$
Equation 4.37

<u>Fractional Area Computations</u>

We may compute the Fractional Partition Areas by dividing each of the component areas by the Area of the Thalean Triangle, A_{ADE}.

In this circumstance A_{ADE} will of course be unity, and the elimination of the dimensional element R will thusly permit the evolution of a grand identity.

The context makes the RHS of Equation 4.17 a convenient divisor:-

$$A_{ADE} = 2R^2 \cos(\alpha) \sin(\alpha)$$
Equation 4.17

and therefore the Fractional Partition Area, F, for Component Area A may formally be stated as:-

$$F = \frac{1}{2R^2 \cos(\alpha) \sin(\alpha)} \cdot A$$
Equation 4.38

The respective Fractional Partition Areas are as follows:-

$$F_{AGB} = \frac{1}{4} \cdot \frac{(\cos(\gamma) - \sin(\gamma))^2}{\cos(\alpha) \cos(\gamma) \cos(\alpha - \gamma)}$$
Equation 4.39

$$F_{FCBG} = \frac{\sin(\gamma)}{2 \cos(\alpha) \cos(\alpha - \gamma)} \left\{ 1 - \tan(\gamma) + \cos(\alpha - \gamma) \left[\frac{(\cos(\alpha) + \sin(\alpha)) \cdot (\cos(\gamma) + \sin(\gamma))}{\cos^2 \alpha - \sin^2(\gamma)} \right] \right\}$$
Equation 4.40

$$F_{EDCF} = \frac{1}{2 \cos(\alpha)} \left[\frac{\sin(\alpha)}{\tan(2\gamma) \cdot \cos(\alpha + \gamma)} + 2 \left\{ \cos(\alpha) - \frac{1}{2} \cdot \frac{\cos(\gamma) + \sin(\gamma)}{\cos(\alpha + \gamma)} \right\} \right]$$
Equation 4.41

Now:-

$$F_{ADE} = F_{AGB} + F_{FCBG} + F_{EDCF}$$
Equation 4.42

or by definition:-

$$1 = F_{AGB} + F_{FCBG} + F_{EDCF}$$
Equation 4.43

Accordingly, by substitution:-

$$1 = \frac{1}{4} \cdot \frac{(\cos(\gamma) - \sin(\gamma))^2}{\cos(\alpha)\cos(\gamma)\cos(\alpha - \gamma)}$$
$$+ \frac{\sin(\gamma)}{2\cos(\alpha)\cos(\alpha - \gamma)}\left\{1 - \tan(\gamma)\right.$$
$$+ \cos(\alpha - \gamma)\left[\frac{(\cos(\alpha) + \sin(\alpha)) \cdot (\cos(\gamma) + \sin(\gamma))}{\cos^2\alpha - \sin^2(\gamma)}\right]\right\}$$
$$+ \frac{1}{2\cos(\alpha)}\left[\frac{\sin(\alpha)}{\tan(2\gamma) \cdot \cos(\alpha + \gamma)} + 2\left\{\cos(\alpha) - \frac{1}{2} \cdot \frac{\cos(\gamma) + \sin(\gamma)}{\cos(\alpha + \gamma)}\right\}\right]$$

Equation 4.44

Simplification of Equation 4.44 enables us to write:-

$$0 = \frac{1}{2\cos(\alpha)} \cdot \left[\frac{1}{2} \cdot \frac{(\cos(\gamma) - \sin(\gamma))^2}{\cos(\gamma).\cos(\alpha - \gamma)}\right.$$
$$+ \frac{\sin(\gamma)}{\cos(\alpha - \gamma)}\left\{1 - \tan(\gamma) + \frac{(\cos(\alpha) + \sin(\alpha)) \cdot (\cos(\gamma) + \sin(\gamma))}{\cos(\alpha + \gamma)}\right\}$$
$$+ \frac{\sin(\gamma)}{\tan(2\gamma) \cdot \cos(\alpha + \gamma)} - \frac{\cos(\gamma) + \sin(\gamma)}{\cos(\alpha + \gamma)}\right]$$

Equation 4.45

Further re-arrangements permit us to formulate a grand identity of quasi-symmetrical structure:-

$$0 \equiv \frac{\sin(\gamma)}{2\cos(\alpha)}\left[\frac{1}{\cos(\alpha - \gamma)}\left\{\frac{1}{\sin(2\gamma)} - \tan(\gamma)\right\} + \frac{(\cos(\alpha) + \sin(\alpha))(\cos(\gamma) + \sin(\gamma))}{\cos(\alpha + \gamma)\cos(\alpha - \gamma)}\right.$$
$$\left. - \frac{1}{\cos(\alpha + \gamma)}\left\{1 + \frac{1}{\sin(2\gamma)}\right\}\right]$$

Equation 4.46

Partitions of the Thalean Triangle:
An Approach via Meister's Rule

by
James R Warren BSc MSc PhD PGCE

The partitions of a triangle may quantitatively be approached by direct inspectional analysis of the geometrical conformation, in practical terms with the assistance of a diagram such as Figure One d. My late disquisitions essayed such avenues. Another approach is to apply Meister's Rule to the analytically-defined co-ordinates of the figure vertices.

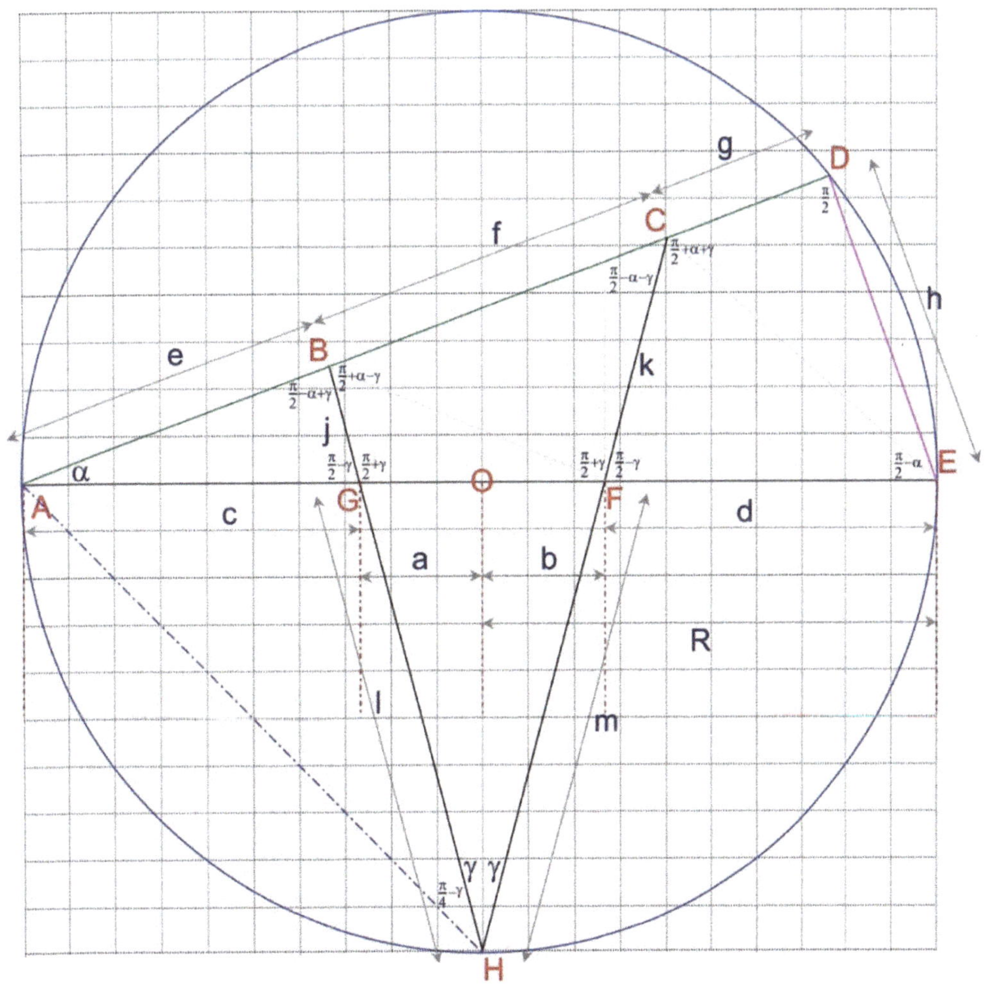

Figure One d
The Geometry of Thalean Triangle Sections from the Nadir

Meister's Rule, sometimes called the Surveyors' Formula, the Shoelace Formula[1] or other things, is an algebraic summation for the area of the general convex-hulled polygon.

Meister's Rule is defined by these alternative equations:-

$$A = \frac{1}{2} \left| \sum_{i=1}^{n-1} x_i y_{i+1} + x_n y_1 - \sum_{i=1}^{n-1} x_{i+1} y_i + x_1 y_n \right|$$

Equation 5.1

which may be elaborated as:-

$$A = \frac{1}{2} \left| x_1 y_2 + x_2 y_3 + \cdots + x_{n-1} y_n + x_n y_1 - x_2 y_1 - x_3 y_2 - \cdots - x_n y_{n-1} - x_1 y_n \right|$$

Equation 5.2

Other forms of Meister's Rule include:-

$$A = \frac{1}{2} \left| \sum_{i=1}^{n} x_i (y_{i+1} - y_{i-1}) \right| = \frac{1}{2} \left| \sum_{i=1}^{n} y_i (x_{i+1} - x_{i-1}) \right| = \frac{1}{2} \left| \sum_{i=1}^{n} x_i y_{i+1} - x_{i+1} y_i \right|$$

Equation 5.3

$$A = \frac{1}{2} \left| \sum_{i=1}^{n} \det \begin{pmatrix} x_i & x_{i+1} \\ y_i & y_{i+1} \end{pmatrix} \right|$$

Equation 5.4

In these equations A is a General Polygon Area; n is the Number of Polygon Vertices (i.e. data point co-ordinates); and x_j and y_j are the respective Abscissal and Ordinal Co-ordinates in two-dimensional Euclidean space.

The summation is positive if points are added in the sinistral sense.

The Meister's Rule summations were computed using both an EXCEL® worksheet and also a MATHCAD® symbolic solution utility: The former for primitives only, and the latter for analytic trigonometric elaborations based upon the co-ordinate primitives. In the following tabulations CIRANG refers to the family of MATHCAD worksheets.

Table One presents EXCEL validation data for a partition system based on nadir sections where Radius R is 10 units; Angle α is 20° and Angle γ is 15°. Note that polygons ADE and EDA, etcetera, are the same figure: EDA is the sinistral, meister-conformable co-ordinate roster.

Polygonal Areas in the Thalean Triangle
By Meister's Rule
consult Figure One

Radius   10

Angle	Diagram Code	Value (degrees)	Value (radians)
α	DAE	20	0.34906585
γ	GHO	15	0.26179939
θ	ABG	85	1.48352986

Line	Value	CIRANG Fiducial	Short Code
AF	12.67949192	12.67949192	12.67949192
j	2.51332518	2.51332518	2.51332518
k	5.29406192	5.29406192	5.29406192
h	6.84040287	6.84040287	6.84040287

Meister Rule Areas

Figure	Area	CIRANG Fiducial
GBA	8.885946399	8.8859464
FCBG	23.53342953	23.53342953
EDCF	31.85938504	31.85938504
ADE		64.27876097
ADE formula	64.27876097	
EDA meister	64.27876097	
ADE sum	64.27876097	

Point	Formula x	y	Value x	y	From Graph x	y
A	-R	0	-10.00000000	0.00000000	-10	0
G	-R.Tanγ	0	-2.67949192	0.00000000	-2.7	0
O	0	0	0.00000000	0.00000000	0	0
F	R.Tanγ	0	2.67949192	0.00000000	2.6	0
E	R	0	10.00000000	0.00000000	10	0
B	-R.Tanγ-j.Sinγ	j.Cosγ	-3.32998835	2.42768570	-3.3	2.45
C	R.Tanγ+k.Sinγ	k.Cosγ	4.04969598	5.11367114	4.04	5.15
D	R-h.Sinα	h.Cosα	7.66044443	6.42787610	7.6	6.45

Table One
Validation Data for Meister's Rule Areal Computation Method

Table Two tabulates the pointwise values of the Meister augends and minuends for each figure point:-

Meister Sums

Figure EDA

Point	Augend	Minuend
E	64.27876097	0.00000000
D	0.00000000	-64.27876097
A	0.00000000	0.00000000
Sum	64.27876097	-64.27876097

Figure GBA

Point	Augend	Minuend
G	-6.50496424	0.00000000
B	0.00000000	-24.27685704
A	0.00000000	0.00000000
G		
Sum	-6.50496424	-24.27685704

Figure FCBG

Point	Augend	Minuend
F	13.70204051	0.00000000
C	9.83138902	-17.02846530
B	0.00000000	-6.50496424
G	0.00000000	0.00000000
Sum	23.53342953	-23.53342953

Figure EDCF

Point	Augend	Minuend
E	64.27876097	0.00000000
D	39.17299357	26.03094396
C	0.00000000	13.70204051
F	0.00000000	0.00000000
Sum	103.45175454	39.73298447

Table Two
Meister's Rule Component Values

There is manifestly no numerical distinction between the classical findings of analytical geometry and the outcome of the Meister's Rule.

<u>Line Segments</u>

The properties of the partitioning system may be described in terms of the Radius of the Circumscriptive Circle, R, and the Angle of Elevation, α, together with the Angle of Section from the Nadir, γ.

In addition, and also in conformity with the geometrical and diagrammatic notations:-

$$h = 2R\sin(\alpha)$$
Equation 5.5

$$j = R(1 - \tan(\gamma))\frac{\sin(\alpha)}{\cos(\alpha - \gamma)}$$
Equation 5.6

$$k = R(1 + \tan(\gamma))\frac{\sin(\alpha)}{\cos(\alpha + \gamma)}$$
Equation 5.7

All three of these sective lines have many identities, but I speculate that the above equations may be the most computationally-efficient, and at least have the virtues of brevity and symmetry.

<u>The Meister Area of Component Triangle GBA</u>

Note these co-ordinates:-

Point	i	x_i	y_i
G	1	$-R.\tan(\gamma)$	0
B	2	$-R.\tan(\gamma) - j.\sin(\gamma)$	$j.\cos(\gamma)$
A	3	$-R$	0

Table Three
Cartesian Co-ordinate Point Equations for Triangle GBA

The elaborate Meister's Rule formula is:-

$$A_{GBA} = \frac{1}{2} \cdot (x_1 y_2 + x_2 y_3 + x_3 y_1 - x_3 y_2 - x_2 y_1 - x_1 y_3)$$
Equation 5.8

Delete null terms A and G:-

$$A_{GBA} = \frac{1}{2} \cdot (x_1 y_2 - x_3 y_2)$$
Equation 5.9

Re-arrange:-

$$A_{GBA} = \frac{1}{2} \cdot \left(y_2 (x_1 - x_3)\right)$$
Equation 5.10

Substitute:-

$$A_{GBA} = \frac{1}{2} \cdot [j.\cos\gamma \cdot (-R\tan(\gamma) - -R)]$$
Equation 5.11

Simplify:-

$$A_{GBA} = \frac{R.\cos(\gamma).(1 - \tan(\gamma))}{2} \cdot j = \frac{R^2}{2} \cdot \frac{\sin(\alpha)\cos(\gamma)(1 - \tan(\gamma))(1 - \tan(\gamma))}{\cos(\alpha - \gamma)}$$
$$= \frac{R^2(1 - \tan(\gamma))^2}{2} \cdot \frac{\sin(\alpha)\cos(\gamma)}{\cos(\alpha - \gamma)}$$
Equation 5.12

<u>The Meister Area of Component Quadrilateral FCBG</u>

Note these co-ordinates:-

Point	i	x_i	y_i
F	1	$R.\tan(\gamma)$	0
C	2	$R.\tan(\gamma)+k.\sin(\gamma)$	$k.\cos(\gamma)$
B	3	$-R.\tan(\gamma)-j.\sin(\gamma)$	$j.\cos(\gamma)$
G	4	$-R.\tan(\gamma)$	0

Table Four
Cartesian Co-ordinate Point Equations for Quadrilateral FCBG

The elaborate Meister's Rule formula is:-

$$A_{FCGB} = \frac{1}{2} \cdot (x_1 y_2 + x_2 y_3 + x_3 y_4 + x_4 y_1 - x_4 y_3 - x_3 y_2 - x_2 y_1 - x_1 y_4)$$

Equation 5.13

Delete null terms F and G:-

$$A_{FCGB} = \frac{1}{2} \cdot (x_1 y_2 + x_2 y_3 - x_4 y_3 - x_3 y_2)$$

Equation 5.14

Re-arrange:-

$$A_{FCGB} = \frac{1}{2} \cdot \left(y_2(x_1 - x_3) + y_3(x_2 - x_4) \right)$$

Equation 5.15

Substitute:-

$$A_{FCGB} = \frac{1}{2} \cdot [k.\cos\gamma\,((R.\tan(\gamma) - (-R.\tan(\gamma) - j.\sin(\gamma)))$$
$$+ j.\cos(\gamma)\,(R.\tan(\gamma) + k.\sin(\gamma) - -R.\tan(\gamma)]$$

Equation 5.16

Condense:-

$$A_{FCGB} = \frac{1}{2} \cdot \left[k.\cos\gamma\,\left(2R.\tan(\gamma) + j.\sin(\gamma)\right) + j.\cos(\gamma)\,\left(2R.\tan(\gamma) + k.\sin(\gamma)\right) \right]$$

Equation 5.17

Abstract cos(γ):-

$$A_{FCGB} = \frac{\cos(\gamma)}{2} \cdot \left[k.\left(2R.\tan(\gamma) + j.\sin(\gamma)\right) + j.\left(2R.\tan(\gamma) + k.\sin(\gamma)\right) \right]$$

Equation 5.18

Develop tan(γ):-

$$A_{FCGB} = \frac{\cos(\gamma)}{2} \cdot \left[k.\left(2R.\tan(\gamma) + j.\cos(\gamma)\tan(\gamma)\right) + j.\left(2R.\tan(\gamma) + k.\cos(\gamma)\tan(\gamma)\right) \right]$$

Equation 5.19

Abstract tan(γ):-

$$A_{FCGB} = \frac{\cos(\gamma).\tan(\gamma)}{2} \cdot \left[k.\left(2R + j.\cos(\gamma)\right) + j.\left(2R + k.\cos(\gamma)\right)\right]$$

Equation 5.20

Make substitutions, install identities and isolate R:-

$$A_{FCGB} = \frac{[R.\sin(\alpha)]^2(1 - \tan(\gamma)^2)}{\cos(\alpha - \gamma).\cos(\alpha + \gamma)}$$
$$\cdot \left[\cos(\gamma)^2.\tan(\gamma) + \frac{\sin(\gamma)}{\sin(\alpha)} \cdot \left(\frac{\cos(\alpha + \gamma)}{1 + \tan(\gamma)} + \frac{\cos(\alpha - \gamma)}{1 - \tan(\gamma)}\right)\right]$$

Equation 5.21

Substitute for $\cos(\gamma)^2.\tan(\gamma)$:-

$$A_{FCGB} = \frac{[R.\sin(\alpha)]^2(1 - \tan(\gamma)^2)}{\cos(\alpha)^2 + \cos(\gamma)^2 - 1} \cdot \left[\frac{\tan(\gamma)}{1 + \tan(\gamma)^2} + \frac{\sin(\gamma)}{\sin(\alpha)} \cdot \left(\frac{\cos(\alpha + \gamma)}{1 + \tan(\gamma)} + \frac{\cos(\alpha - \gamma)}{1 - \tan(\gamma)}\right)\right]$$

Equation 5.22

Substitute for $\tan(\gamma)/(1+\tan(\gamma)^2)$:-

$$A_{FCGB} = \frac{[R.\sin(\alpha)]^2(1 - \tan(\gamma)^2)}{\cos(\alpha)^2 + \cos(\gamma)^2 - 1} \cdot \left[\sin(\gamma).\cos(\gamma) + \frac{\sin(\gamma)}{\sin(\alpha)} \cdot \left(\frac{\cos(\alpha + \gamma)}{1 + \tan(\gamma)} + \frac{\cos(\alpha - \gamma)}{1 - \tan(\gamma)}\right)\right]$$

Equation 5.23

Isolate $\sin(\gamma)$:-

$$A_{FCGB} = \frac{[R.\sin(\alpha)]^2.\sin(\gamma).(1 - \tan(\gamma)^2)}{\cos(\alpha)^2 + \cos(\gamma)^2 - 1}$$
$$\cdot \left[\cos(\gamma) + \frac{1}{\sin(\alpha)} \cdot \left(\frac{\cos(\alpha + \gamma)}{1 + \tan(\gamma)} + \frac{\cos(\alpha - \gamma)}{1 - \tan(\gamma)}\right)\right]$$

Equation 5.24

Eliminate $(1-\tan(\gamma)^2)$:-

$$A_{FCGB} = \frac{2.[R.\sin(\alpha)]^2.\sin(\gamma).\tan(\gamma)}{\tan(2\gamma).[\cos(\alpha)^2 + \cos(\gamma)^2 - 1]} \cdot \left[\cos(\gamma) + \frac{1}{\sin(\alpha)} \cdot \left(\frac{\cos(\alpha + \gamma)}{1 + \tan(\gamma)} + \frac{\cos(\alpha - \gamma)}{1 - \tan(\gamma)}\right)\right]$$

Equation 5.25

Substitute for $\sin(\alpha)^2$:-

$$A_{FCGB} = \frac{R^2(1 - \cos(2\alpha)).\sin(\gamma).\tan(\gamma)}{\tan(2\gamma).[\cos(\alpha)^2 + \cos(\gamma)^2 - 1]} \cdot \left[\cos(\gamma) + \frac{1}{\sin(\alpha)} \cdot \left(\frac{\cos(\alpha+\gamma)}{1+\tan(\gamma)} + \frac{\cos(\alpha-\gamma)}{1-\tan(\gamma)}\right)\right]$$

Equation 5.26

Substitute for $\cos(\alpha)^2$ and $\cos(\gamma)^2$:-

$$A_{FCGB} = \frac{R^2(1 - \cos(2\alpha)).\sin(\gamma).\tan(\gamma)}{\tan(2\gamma).\left[\dfrac{\cos(2\alpha) + \cos(2\gamma) + 2}{2} - 1\right]}$$
$$\cdot \left[\cos(\gamma) + \frac{1}{\sin(\alpha)} \cdot \left(\frac{\cos(\alpha+\gamma)}{1+\tan(\gamma)} + \frac{\cos(\alpha-\gamma)}{1-\tan(\gamma)}\right)\right]$$

Equation 5.27

Remove +2 and -1:-

$$A_{FCGB} = \frac{R^2(1 - \cos(2\alpha)).\sin(\gamma).\tan(\gamma)}{\tan(2\gamma).\left[\dfrac{\cos(2\alpha) + \cos(2\gamma)}{2}\right]} \cdot \left[\cos(\gamma) + \frac{1}{\sin(\alpha)} \cdot \left(\frac{\cos(\alpha+\gamma)}{1+\tan(\gamma)} + \frac{\cos(\alpha-\gamma)}{1-\tan(\gamma)}\right)\right]$$

Equation 5.28

Re-arrange 2 as a divisor:-

$$A_{FCGB} = \frac{2.R^2(1 - \cos(2\alpha)).\sin(\gamma).\tan(\gamma)}{\tan(2\gamma).[\cos(2\alpha) + \cos(2\gamma)]}$$
$$\cdot \left[\cos(\gamma) + \frac{1}{\sin(\alpha)} \cdot \left(\frac{\cos(\alpha+\gamma)}{1+\tan(\gamma)} + \frac{\cos(\alpha-\gamma)}{1-\tan(\gamma)}\right)\right]$$

Equation 5.29

Simplify:-

$$A_{FCGB} = \frac{2.R^2.\sin(2\alpha).\tan(\alpha).\sin(\gamma).\tan(\gamma)}{\tan(2\gamma).[\cos(2\alpha) + \cos(2\gamma)]}$$
$$\cdot \left[\cos(\gamma) + \frac{1}{\sin(\alpha)} \cdot \left(\frac{\cos(\alpha+\gamma)}{1+\tan(\gamma)} + \frac{\cos(\alpha-\gamma)}{1-\tan(\gamma)}\right)\right]$$

Equation 5.30

Substitute for $\tan(\gamma)$ and simplify:-

$$A_{FCGB} = \frac{2.R^2.\sin(2\alpha).\tan(\alpha).\sin(\gamma)^2}{\tan(2\gamma).[\cos(2\alpha) + \cos(2\gamma)]} \cdot \left[1 + \frac{1}{\sin(\alpha).\cos(\gamma)} \cdot \left(\frac{\cos(\alpha+\gamma)}{1+\tan(\gamma)} + \frac{\cos(\alpha-\gamma)}{1-\tan(\gamma)}\right)\right]$$

Equation 5.31

Substitute for sin(2α) and simplify:-

$$A_{FCGB} = \frac{2^2.R^2.\sin^2(\alpha).\sin^2(\gamma)}{\tan(2\gamma).[\cos(2\alpha) + \cos(2\gamma)]} \cdot \left[1 + \frac{1}{\sin(\alpha).\cos(\gamma)} \cdot \left(\frac{\cos(\alpha+\gamma)}{1+\tan(\gamma)} + \frac{\cos(\alpha-\gamma)}{1-\tan(\gamma)}\right)\right]$$

Equation 5.32

Consolidate squares:-

$$A_{FCGB} = \frac{[2R\sin(\alpha)\sin(\gamma)]^2}{\tan(2\gamma).[\cos(2\alpha) + \cos(2\gamma)]} \cdot \left[1 + \frac{1}{\sin(\alpha).\cos(\gamma)} \cdot \left(\frac{\cos(\alpha+\gamma)}{1+\tan(\gamma)} + \frac{\cos(\alpha-\gamma)}{1-\tan(\gamma)}\right)\right]$$

Equation 5.33

<u>The Meister Area of Component Quadrilateral EDCF</u>

Note these co-ordinates:-

Point	i	x_i	y_i
E	1	R	0
D	2	R-h.sin(α)	h.cos(α)
C	3	R.tan(γ)+k.sin(γ)	k.cos(γ)
F	4	R.tan(γ)	0

Table Five
Cartesian Co-ordinate Point Equations for Quadrilateral EDCF

The elaborate Meister's Rule formula is:-

$$A_{EDCF} = \frac{1}{2} \cdot (x_1 y_2 + x_2 y_3 + x_3 y_4 + x_4 y_1 - x_4 y_3 - x_3 y_2 - x_2 y_1 - x_1 y_4)$$

Equation 5.34

Substitute formulae for co-ordinates:-

$$A_{EDCF} = \frac{1}{2} \cdot (R.h.\cos(\alpha) + (R - h.\sin(\alpha)).k.\cos(\gamma) + (R.\tan(\gamma) + k.\sin(\gamma)).0$$
$$+ R.\tan(\gamma).0 - R.\tan(\gamma).k.\cos(\gamma) - (R.\tan(\gamma) + k.\sin(\gamma)).h.\cos(\alpha)$$
$$- (R - h.\sin(\alpha)).0 - R.0)$$

Equation 5.35

Remove null terms:-

$$A_{EDCF} = \frac{1}{2} \cdot (R.h.\cos(\alpha) + (R - h.\sin(\alpha)).k.\cos(\gamma) - R.\tan(\gamma).k.\cos(\gamma)$$
$$- (R.\tan(\gamma) + k.\sin(\gamma)).h.\cos(\alpha))$$

Equation 5.36

Expand internally:-

$$A_{EDCF} = \frac{1}{2} \cdot (R.h.\cos(\alpha) + R.k.\cos(\gamma) - h.k.\sin(\alpha).\cos(\gamma) - R.k.\tan(\gamma).\cos(\gamma)$$
$$- R.h.\cos(\alpha).\tan(\gamma) - k.h.\cos(\alpha).\sin(\gamma))$$

Equation 5.37

Note that:-

$$k = \frac{h.(1 + \tan(\gamma))}{2.\cos(\alpha + \gamma)}$$

Equation 5.38

Substitute Equation 5.38 for k in Equation 5.37:-

$$A_{EDCF} = \frac{1}{2} \cdot \left(R.h.\cos(\alpha) + R.\frac{h.(1 + \tan(\gamma))}{2.\cos(\alpha + \gamma)}.\cos(\gamma) - h.\frac{h.(1 + \tan(\gamma))}{2.\cos(\alpha + \gamma)}.\sin(\alpha).\cos(\gamma) \right.$$
$$- R.\frac{h.(1 + \tan(\gamma))}{2.\cos(\alpha + \gamma)}.\tan(\gamma).\cos(\gamma)$$
$$\left. - R.h.\cos(\alpha).\tan(\gamma) - \frac{h.(1 + \tan(\gamma))}{2.\cos(\alpha + \gamma)}.h.\cos(\alpha).\sin(\gamma) \right)$$

Equation 5.39

Resolve terms in Rh and h^2:-

$$A'_{EDCF} = \frac{1}{2}\left(R.h.\left\{ \cos(\alpha) + \frac{(1+\tan(\gamma))}{2.\cos(\alpha+\gamma)}.\cos(\gamma) \right.\right.$$
$$\left. - \frac{(1+\tan(\gamma))}{2.\cos(\alpha+\gamma)}.\tan(\gamma).\cos(\gamma) - \cos(\alpha).\tan(\gamma) \right\}$$
$$\left. - h^2\left\{ \frac{(1+\tan(\gamma))}{2.\cos(\alpha+\gamma)}.\sin(\alpha).\cos(\gamma) + \frac{(1+\tan(\gamma))}{2.\cos(\alpha+\gamma)}.\cos(\alpha).\sin(\gamma) \right\}\right)$$

Equation 5.40

Re-arrange for $(1-\tan(\gamma))$:-

$$A_{EDCF} = \frac{1}{2}\left(R.h.\left\{ \frac{(1+\tan(\gamma))}{2.\cos(\alpha+\gamma)}.\cos(\gamma).(1-\tan(\gamma)) + \cos(\alpha).(1-\tan(\gamma)) \right\}\right.$$
$$\left. - h^2\left\{ \frac{(1+\tan(\gamma))}{2.\cos(\alpha+\gamma)}.\sin(\alpha).\cos(\gamma) + \frac{(1+\tan(\gamma))}{2.\cos(\alpha+\gamma)}.\cos(\alpha).\sin(\gamma) \right\}\right)$$

Equation 5.41

Re-arrange:-

$$A_{EDCF} = \frac{1}{2}\left(R.h.(1-\tan(\gamma))\left\{ \frac{(1+\tan(\gamma))}{2.\cos(\alpha+\gamma)}.\cos(\gamma) + \cos(\alpha)) \right\}\right.$$
$$\left. - h^2.\frac{(1+\tan(\gamma))}{2.\cos(\alpha+\gamma)}.\{\sin(\alpha).\cos(\gamma) + \cos(\alpha).\sin(\gamma)\} \right)$$

Equation 5.42

Substitute Sine Sum Identity:-

$$A_{EDCF} = \frac{1}{2}\left(R.h.(1-\tan(\gamma))\left\{ \frac{(1+\tan(\gamma))}{2.\cos(\alpha+\gamma)}.\cos(\gamma) + \cos(\alpha)) \right\}\right.$$
$$\left. - h^2.\frac{(1+\tan(\gamma))}{2.\cos(\alpha+\gamma)}.\{\sin(\alpha+\gamma\} \right)$$

Equation 5.43

Substitute for h:-

$$A_{EDCF} = \frac{1}{2}\left(R.\,2R\sin(\alpha).(1-\tan(\gamma))\left\{ \frac{(1+\tan(\gamma))}{2.\cos(\alpha+\gamma)}.\cos(\gamma)+\cos(\alpha)) \right\} \right.$$
$$\left. -\,4R^2\sin(\alpha)^2.\frac{(1+\tan(\gamma))}{2.\cos(\alpha+\gamma)}.\{\sin(\alpha+\gamma\} \right)$$

Equation 5.44

Simplify (the tactical substitution of tan(α+γ) has been deferred):-

$$A_{EDCF} = R^2\left(\sin(\alpha).(1-\tan(\gamma))\left\{ \frac{1+\tan(\gamma)}{2.\cos(\alpha+\gamma)}.\cos(\gamma)+\cos(\alpha)) \right\} \right.$$
$$\left. -\,\sin(\alpha)^2.\frac{(1+\tan(\gamma))}{\cos(\alpha+\gamma)}.\{\sin(\alpha+\gamma\} \right)$$

Equation 5.45

Abstract sin(α):-

$$A_{EDCF} = R^2.\sin(\alpha).\left((1-\tan(\gamma))\left\{ \frac{1+\tan(\gamma)}{2.\cos(\alpha+\gamma)}.\cos(\gamma)+\cos(\alpha)) \right\} \right.$$
$$\left. -\,\sin(\alpha).\frac{1+\tan(\gamma)}{\cos(\alpha+\gamma)}.\sin(\alpha+\gamma) \right)$$

Equation 5.46

Multiply (1-tan(γ)) and (1+tan(γ)):-

$$A_{EDCF} = R^2.\sin(\alpha).\left(\left\{ \frac{(1-\tan(\gamma))\cdot(1+\tan(\gamma))}{2.\cos(\alpha+\gamma)}.\cos(\gamma)+\cos(\alpha).(1-\tan(\gamma)) \right\} \right.$$
$$\left. -\,\sin(\alpha).\frac{1+\tan(\gamma)}{\cos(\alpha+\gamma)}.\sin(\alpha+\gamma) \right)$$

Equation 5.47

Simplify:-

$$A_{EDCF} = R^2.\sin(\alpha).\left(\left\{ \frac{\cos^2(\gamma)-\sin^2(\gamma)}{2.\cos(\gamma).\cos(\alpha+\gamma)}+\cos(\alpha).\frac{\cos(\gamma)-\sin(\gamma)}{\cos(\gamma)} \right\} \right.$$
$$\left. -\,\frac{\sin(\alpha)}{\cos(\gamma)}.\frac{\cos(\gamma)+\sin(\gamma)}{\cos(\alpha+\gamma)}.\sin(\alpha+\gamma) \right)$$

Equation 5.48

Simplify brackets and abstract cos(γ):-

$$A_{EDCF} = \frac{R^2 \sin(\alpha)}{\cos(\gamma)} \cdot \left(\frac{\cos^2(\gamma) - \sin^2(\gamma)}{2.\cos(\alpha + \gamma)} + \cos(\alpha).(\cos(\gamma) - \sin(\gamma)) \right.$$
$$\left. - \frac{\cos(\gamma) + \sin(\gamma)}{\cos(\alpha + \gamma)} . \sin(\alpha) . \sin(\alpha + \gamma) \right)$$

Equation 5.49

Simplify:-

$$A_{EDCF} = \frac{R^2 \sin(\alpha)}{\cos(\gamma)} \cdot \left(\frac{(\cos(\gamma) + \sin(\gamma)) \cdot (\cos(\gamma) - \sin(\gamma))}{2.\cos(\alpha + \gamma)} + \cos(\alpha).(\cos(\gamma) - \sin(\gamma)) \right.$$
$$\left. - \frac{\cos(\gamma) + \sin(\gamma)}{\cos(\alpha + \gamma)} . \sin(\alpha) . \sin(\alpha + \gamma) \right)$$

Equation 5.50

Abstract (cos(γ)+sin(γ)):-

$$A_{EDCF} = \frac{R^2 \sin(\alpha)(\cos(\gamma) + \sin(\gamma))}{\cos(\gamma)} \cdot \left(\frac{\cos(\gamma) - \sin(\gamma)}{2.\cos(\alpha + \gamma)} + \cos(\alpha).\frac{\cos(\gamma) - \sin(\gamma)}{\cos(\gamma) + \sin(\gamma)} \right.$$
$$\left. - \frac{\sin(\alpha).\sin(\alpha + \gamma)}{\cos(\alpha + \gamma)} \right)$$

Equation 5.51

Abstract (cos(γ)-sin(γ)):-

$$A_{EDCF} = \frac{R^2 \sin(\alpha)(\cos(\gamma) + \sin(\gamma)) (\cos(\gamma) - \sin(\gamma))}{\cos(\gamma)} \cdot \left(\frac{1}{2.\cos(\alpha + \gamma)} \right.$$
$$\left. + \frac{\cos(\alpha)}{\cos(\gamma) + \sin(\gamma)} - \frac{\sin(\alpha).\tan(\alpha + \gamma)}{\cos(\gamma) - \sin(\gamma)} \right)$$

Equation 5.52

Re-arrange:-

$$A_{EDCF} = \frac{R^2 \sin(\alpha)(\cos^2(\gamma) - \sin^2(\gamma))}{\cos(\gamma)} \cdot \left(\frac{1}{2 \cdot \cos(\alpha + \gamma)} + \frac{\cos(\alpha)}{\cos(\gamma) + \sin(\gamma)} - \frac{\sin(\alpha) \cdot \tan(\alpha + \gamma)}{\cos(\gamma) - \sin(\gamma)} \right)$$

Equation 5.53

Note the identity:-

$$\frac{\cos^2(\gamma) - \sin^2(\gamma)}{\cos(\gamma)} \equiv \frac{2\tan(\gamma)}{\tan(2\gamma)}$$

Equation 5.54

Substitute the identity and simplify:-

$$A_{EDCF} = \frac{2R^2 \sin(\alpha)\sin(\gamma)}{\tan(2\gamma)} \cdot \left(\frac{1}{2 \cdot \cos(\alpha + \gamma)} + \frac{\cos(\alpha)}{\cos(\gamma) + \sin(\gamma)} - \frac{\sin(\alpha) \cdot \tan(\alpha + \gamma)}{\cos(\gamma) - \sin(\gamma)} \right)$$

Equation 5.55

On my MATHCAD scratchpad a further ten pages of intricate substitutions and expansions followed Equation 5.55 but the vast majority of these were prolix and profitless.

The exceptions were the following forms which may be as or more economical than Equation 5.55, or at any event share with it the virtues of concision and symmetry.

<u>An Alternate Form for the Area of Quadrilateral EDCF</u>

This alternative form arose from extended derivations from Equation 5.55, and expresses parts of the area in terms of the functions of double angles:-

$$A_{EDCF} = \frac{R^2 \sin(\alpha)}{\cos(\alpha + \gamma)} \cdot \left[\cos(\gamma) - \frac{1}{2\cos(\gamma)} + \cos(2\alpha).\cos(\gamma) - \sin(\gamma).(\sin(2\alpha) + 1)\right]$$

Equation 5.56

Develop $\cos(\gamma)$:-

$$A_{EDCF} = \frac{R^2 \sin(\alpha)}{\cos(\alpha + \gamma)}$$
$$\cdot \left[\cos(\gamma) - \frac{1}{2\cos(\gamma)} + \cos(2\alpha).\cos(\gamma) - \cos(\gamma).\tan(\gamma).(\sin(2\alpha) + 1)\right]$$

Equation 5.57

Abstract $\cos(\gamma)$:-

$$A_{EDCF} = \frac{R^2 \sin(\alpha).\cos(\gamma)}{\cos(\alpha + \gamma)} \cdot \left[1 - \frac{1}{2\cos(\gamma)^2} + \cos(2\alpha) - \tan(\gamma).(\sin(2\alpha) + 1)\right]$$

Equation 5.58

Re-arrange:-

$$A_{EDCF} = \frac{R^2 \sin(\alpha).\cos(\gamma)}{\cos(\alpha + \gamma)} \cdot \left[1 + \cos(2\alpha) - \tan(\gamma).(\sin(2\alpha) + 1) - \frac{1}{2\cos(\gamma)^2}\right]$$

Equation 5.59

Note the identity:-

$$1 + \cos(2\alpha) \equiv 2\cos^2(\alpha)$$
Equation 5.60

Substitute the identity in Equation 5.59:-

$$A_{EDCF} = \frac{R^2 \sin(\alpha).\cos(\gamma)}{\cos(\alpha + \gamma)} \cdot \left[2\cos^2(\alpha) - \tan(\gamma).(\sin(2\alpha) + 1) - \frac{1}{2\cos(\gamma)^2} \right]$$

Equation 5.61

<u>Area Check</u>

The Total Area of the Thalean Triangle ADE is given by:-

$$A_{ADE} = 2R^2 \cos(\alpha) \sin(\alpha)$$
Equation 5.62

There are of course an infinite number of equations for the area of any triangle but I have found the form of Equation 5.62 convenient for normalising the partition areas of the thalean triangle.

For the given values of R, α and γ Equation 5.62 computes to 64.2787609686539 units.

It also of course the case that:-

$$A_{ADE} = A_{GBA} + A_{FCGB} + A_{EDCF}$$
Equation 5.63

The value yielded by Equation 5.63 also proved to be 64.2787609686539 units.

<u>Fractional Area Computations</u>

The Fractional Area of a Partition is defined as:-

$$F_A = \frac{A_A}{A_{ADE}} = \frac{A_A}{2R^2 \cos(\alpha) \sin(\alpha)}$$
Equation 5.64

where A_A is the Area of a Partition, i.e. A_{GBA}, A_{FCGB} or A_{EDCF}. Accordingly:-

$$1 = F_{GBA} + F_{FCGB} + F_{EDCF}$$
Equation 5.65

whilst:-

$$F_{GBA} = \frac{1}{4} \cdot \frac{\cos(\gamma) \cdot (1 - \tan(\gamma))^2}{\cos(\alpha) \cdot \cos(\alpha - \gamma)}$$

Equation 5.66

$$F_{FCGB} = \frac{2 \tan(\alpha) \sin^2(\gamma)}{\tan(2\gamma) \cdot [\cos(2\alpha) + \cos(2\gamma)]} \cdot \left[1 + \frac{1}{\sin(\alpha) \cdot \cos(\gamma)} \cdot \left(\frac{\cos(\alpha + \gamma)}{1 + \tan(\gamma)} + \frac{\cos(\alpha - \gamma)}{1 - \tan(\gamma)}\right)\right]$$

Equation 5.67

and:-

$$F_{EDCF} = \frac{\sin(\gamma)}{\cos(\alpha) \cdot \tan(2\gamma)} \cdot \left(\frac{1}{2 \cdot \cos(\alpha + \gamma)} + \frac{\cos(\alpha)}{\cos(\gamma) + \sin(\gamma)} - \frac{\sin(\alpha) \cdot \tan(\alpha + \gamma)}{\cos(\gamma) - \sin(\gamma)}\right)$$

Equation 5.68

By substituting Equations 5.66, 5.67 and 5.68 in Equation 5.65 it is possible to establish the grand identity:-

$$1 \equiv \frac{1}{4} \cdot \frac{\cos(\gamma) \cdot (1 - \tan(\gamma))^2}{\cos(\alpha) \cdot \cos(\alpha - \gamma)} + \frac{2 \tan(\alpha) \sin^2(\gamma)}{\tan(2\gamma) \cdot [\cos(2\alpha) + \cos(2\gamma)]}$$
$$\cdot \left[1 + \frac{1}{\sin(\alpha) \cdot \cos(\gamma)} \cdot \left(\frac{\cos(\alpha + \gamma)}{1 + \tan(\gamma)} + \frac{\cos(\alpha - \gamma)}{1 - \tan(\gamma)}\right)\right]$$
$$+ \frac{\sin(\gamma)}{\cos(\alpha) \cdot \tan(2\gamma)} \cdot \left(\frac{1}{2 \cdot \cos(\alpha + \gamma)} + \frac{\cos(\alpha)}{\cos(\gamma) + \sin(\gamma)} - \frac{\sin(\alpha) \cdot \tan(\alpha + \gamma)}{\cos(\gamma) - \sin(\gamma)}\right)$$

Equation 5.69

CHAPTER FOUR

Examinations of
Barr's Buffon Volume Ten

Barr's Buffon
and the
Loss of Light

by
James R Warren BSc MSc PhD PGCE

A man that looks on glass
On it may stay his eye;
Or if he pleaseth, through it pass,
And then the heaven espy

- George Herbert

Science is an art of some subtlety. It is difficult for an environmental scientist to negotiate physical science, or for a natural philosopher to expound natural history. And presiding over all endeavors is that jesting genius mathematics, pure and sophistical, essential but reclusive, charming and treacherous. I know this because as an old geologist I have tried many trespasses.

This is easy for me to say. For Herbert, writing in 1633, at the end of his short life, the glass was more obscure and the World of our Father clearer. As soon as men invoked technique and betrothed it to wisdom, then a stormy marriage was made. Such is enlightenment.

In terms of classical wave theory the energy density of a wave, or in other words its power exerted through a given volume or over a unit area, relates proportionately to the square of the wave's amplitude.

The details of this transfer can be complex, in both the mathematical and the colloquial senses.

The positon is further involved by the fact that waves enfeeble geometrically as they spread from their source into ever-dilating spheres or circles; and enfeeble additionally due to impediments from whatever resistances, if any, through which they expand. And when we investigate energy flux at astronomical or atomic scales Classical concepts like "sphere" and "circle" lose their meanings and time and change lose all mortal resolution, or perhaps their resolutions are only all too mortal?

Investigators of the Heroic Ages of science knew as if by instinct of these issues, and sought to mitigate them by the cunning contrivance of their experiments. For example, some experimenters sought to marshal or corral the waves of interest to them by training them through mirrors, lenses or prisms, or, crucially, by collimating the energy through gaps and shutters. This could be done not only with light, but with more or less success in the case of sound or surficial gravity waves.

The geometrical and material attenuations of energy were, and are, a source of confusion to many, even to men of the stature of Georges-Louis Leclerc, First Comte de Buffon (1707-1788), a very great mathematician, scientist and naturalist who gave us thirty-six volumes of his seminal observations upon the manifest and recondite about us. As you possibly surmise, the confounding was compounded when the hapless Barr[1] attempted to translate the wisdom of the foreign savant into our own language, and when this monoglot tried to understand the words of both.

I rather think that it was a crisp spring day in 2014 when I walked up Dartmouth Park Hill to Highgate Village. I had little purpose in mind but happened into the Highgate Bookshop and browsed for whatever might have interested me. A battered ancient volume of leather-bound letterpress caught my eye as it sat forlornly amidst more modern books on a shelf beneath the counter. It was Volume Ten of Barr's English translation of Buffon's *Histoire Naturelle*, bereft of its thirty-five sisters. The frontispiece is reproduced in Appendix One.

I opened the book and was immediately intrigued by a list of figures:-

Thicknefs	1	2	3	4	5	6
Diminution	$\dfrac{2}{7}$	$\dfrac{10}{47}$	$\dfrac{50}{343}$	$\dfrac{250}{2401}$	$\dfrac{1250}{16807}$	$\dfrac{6250}{117649}$

Table One
Bouguer's Alleged Pane Attenuation Fractions
As stated in Barr's Buffon Volume Ten

The archaic list of vulgar fractions, and their unfamiliar precision provoked me with their obvious potential for analysis: The very precision betokened dubiety. After all, decimals have authority. And it was clear from the text which followed that Buffon too had had his doubts, and had sought in 1747 to confirm or deny these results established eighteen years earlier by the pioneering geophysicist and hydrographer Pierre Bouguer (1698-1758).[2]

But I am a poor man. Highgate is an expensive place. The asking price for the book, pencilled on the fly leaf, was £25. I loitered for some time around the shelves in a state of perplexed indecision. Then I decided. I bought the book and scuttled back to my Wife's sister's house before further temptation beggared me.

I reproduce the most relevant passages of Buffon's replicative thinking in the facsimiles of Appendix Two.

It is clear that even world-historical geniuses could be confused by the issues at play here, and it occurred to me that Buffon had probably made a blameless hash of his experiment and his criticism of Bouguer, but of course I could throw no light on the matter (no pun intended, and I trust none taken) unless I performed at least the basic analyses of which I am able.

So what occasions the loss of light?

As remarked above, Intensity is a measure of energy transfer in light. We can specify additive components of Intensity as:-

$$I = R + A + T + S$$
Equation 1

where I is Intensity of Light ("brightness"); R is amount of Intensity Reflected as the light negotiates the front and rear surfaces of the glass pane (or whatever else it is shining through); A is the Loss of Intensity due to the Absorption of Light by Material particles whose size is comparable to the wavelength of the light in question (i.e. molecules or atoms); T is the Light Transmitted through the medium (which greatly interests us); and finally S is the Loss of Light due to Scatter. Scatter is where waves are dispersed at random like a scared flock. This is due to macroscopic particles like dust motes, smoke particles and general dirt. It includes bubbles and flaws in glass, and, as a practical matter, its bulk color (pedantically, an absorptive issue). All these are problems which would have been significant to pre-industrial experimentalists, dependent as they were upon the sources of their materials, and the personal skill of craftsmen.

Therefore, it is pertinent to address the general filth of the experimental materials and the cleanliness or otherwise of the air through which the energy is shone, and Buffon realised this.

His battery of six panes comprised glass sheets each of a thickness of one Royal *ligne*, separated by air gaps of one Royal *pouce*. There are twelve pouces in one Royal French foot, the *pied du roi*. A *pied du roi* is about 0.3248 meters. Further, there are twelve lignes (lines) in one *pouce*. It follows that a *ligne* is 1/144 of a *pied,* or roughly 0.002256 meters. We may also consider the *toise*, or French fathom, which was six feet, or approximately 1.949 meters. These measures have the same names and ratios as the old British Imperial system (a curse of my boyhood) but are slightly bigger.[3]

Because both Bouguer and Buffon were men of quality we can assume that they used Royal measure, but virtually every town of France had its own standards, so exact understanding of ancient work is possibly irrecoverable.

When I recalculated Bouguer's results for myself I assumed of course incorrectly that he used excellent American twenty-first century glass, rolled and polished with micrometer precision, and I attempted to simulate a dusty laboratory in *ancien régime* Paris using an atmospheric absorption coefficient computed for modern heavily polluted conditions.

It is not clear to me whether Bouguer separated his glass elements with air gaps, or if so their size. Because of the relative perspicacity of even dirty air it is difficult to tell from his table of figures, but the implication seems to me that he worked with panes stacked in contact.

Having briefly touched some of the practical issues we can now arrange Equation One for the desideratum, T:-

$$T = I_0 - (R + A + S)$$
Equation 2

where I_0 is the Incident Intensity of the radiation. I will compute R and A at each pane and air gap, but sadly I shall neglect scatter, which is nevertheless important when we are discussing other than modern clean rooms and fine analytic materials.

<u>Light Intensity as a Metric of Imparted Energy</u>

In the context of the Bouguer-Buffon experiments, Light Intensity is the quotient of Power divided by Area:-

$$I = \frac{Power}{Area} = \frac{Energy}{Area \cdot Time}$$

Equation 3

In Classical Mechanics the General Integral for the definition of Energy[4] is:-

$$\int_0^A F(x).\,dx = \int_0^A kx.\,dx = \frac{1}{2} \cdot kA^2$$

Equation 4

where A is Wave Amplitude and F(x) is an appropriate function. Do not think that k is a simple constant. It is a compound term assembled of constants and variables germane to the context.

In the classical treatment of optical wave energy due to James Clerk-Maxwell and his Victorian collaborators we may state in particular that:-

$$I = \frac{cn\epsilon_0}{2} \cdot |E|^2$$

Equation 5

where c is the Celerity of Light (299792458 m/s); n is the Refractive Index of the Material (basically the ratio of the Celerity of Light in Free Space to its speed in the medium concerned); ε_0 is the Permittivity of Free Space and E is the Complex Electric Field Amplitude. ε_0, the Permittivity of Free Space, is about $8.854187817 \times 10^{-12}$ farads per meter. The Refractive Index can be expected to be around 1.5 for artificial glass. The Refractive Index, n, is dimensionless.

Therefore, I and E are intercomputable for a given material.

<u>The Attenuation of Intensity</u>

It is often convenient to express the enfeeblement or attenuation of Intensity by reference to an original input Intensity I_0 which is taken to be unity, from which it follows that:-

$$I_x = I_0(1 - \xi)$$
Equation 6

where I_x is the local or Exiting Intensity and ξ is some dimensionless factor that combines the quantified wave weakening influences. ξ lies between zero and unity. If $\xi=1$ then extinction occurs, or in other words the radiation is totally blocked.

It follows that the quotient I_x/I_0 is itself often a convenient dimensionless metric of attenuation.

<u>Refraction</u>

When a wave penetrates a material surface it experiences refraction. Refraction is a sharp deviation in the straight-line trajectory of the wave path at the surface, and is due to the differential wave celerities within and outwith the medium.

It is because of refraction that lenses may be used to control light.

It is convenient to discuss refraction in terms of the Angle of Incidence θ_i and the Angle of Transmission θ_t. The Angle of Incidence is made to the normal to the surface of the transmitting medium in that medium (usually air) that is travelled *before* refraction. The Angle of Transmission (Transmittance) is made to the normal *within* the second medium (typically glass).

Furthermore, the Refractive Index of the originating medium (air) is n_1 and that of the transmissive medium (glass) is n_2.

θ_i, θ_t, n_1 and n_2 are related by Snell's Law:-

$$\frac{\sin \theta_i}{\sin \theta_t} = \frac{v_1}{v_2} = \frac{\lambda_1}{\lambda_2} = \frac{n_2}{n_1}$$
Equation 7

where v are the respective Phase Velocities (i.e. wave celerities) and λ are the respective Wavelengths.

The geometry of the refraction is perhaps clarified by Figure One, where inspection readily discloses these facts:-

$$\sin \theta_i = \frac{3}{5} \qquad \sin \theta_t = \frac{2}{\sqrt{40}}$$

$$\frac{\sin\theta_i}{\sin\theta_t} = \frac{v_1}{v_2} = \frac{\lambda_1}{\lambda_2} = \frac{n_2}{n_1} = \frac{n_2}{1} = \frac{\frac{3}{5}}{\frac{2}{\sqrt{40}}}$$

From which it is apparent that $\theta_i = 36.869898$ degrees; $\theta_t = 18.434949$ degrees and:-

$$n_2 = \frac{3}{5} \times \frac{\sqrt{40}}{2} = \frac{3\sqrt{40}}{10} = 1.8973665961$$

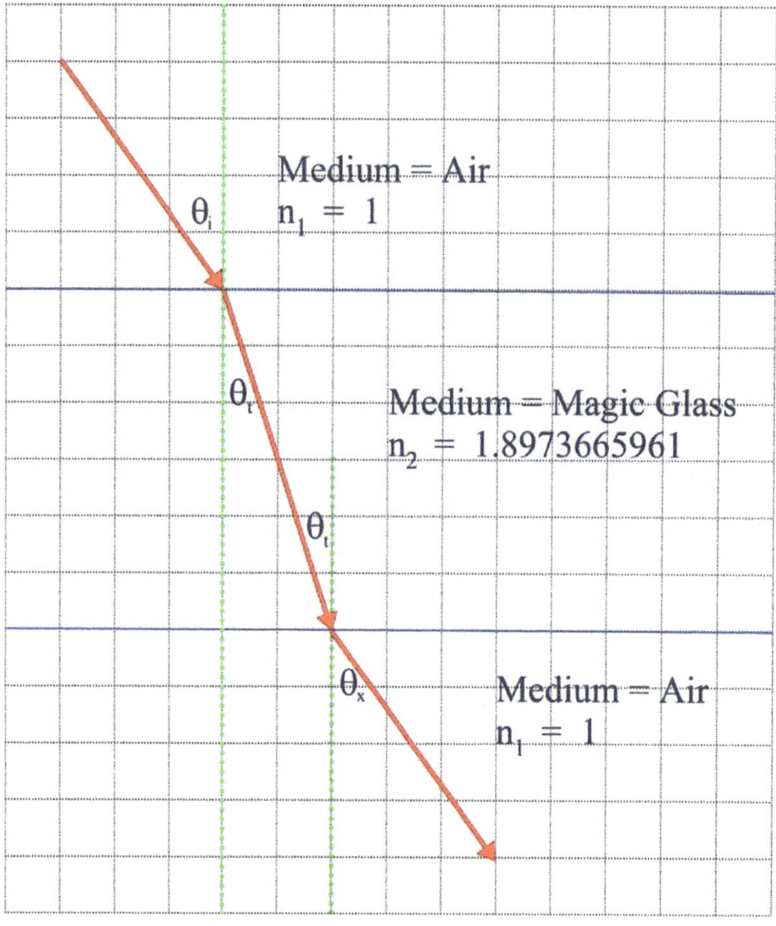

Figure One
The Geometry of Light Wave Refraction

The trigonometry confirms that:-

$$\theta_i = \sin^{-1}\left(\frac{n_1}{n_2} \cdot \sin\theta_t\right)$$
Equation 8

and:-

$$\theta_x = \sin^{-1}\left(\frac{n_2}{n_1} \cdot \sin\theta_t\right)$$
Equation 9

From which it is clear that:-

$$\theta_x = \theta_i$$
Equation 10

Obviously, Equation Ten assumes that the radiation re-enters its original medium, as in the case air to glass and glass to air.

<u>Losses due to Reflection</u>

The Reflectance, R, quantifies the fraction of radiation sent back from each interface. Therefore, we must assess R for both the leading and trailing surfaces of a pane.

Furthermore, reflected light consists of both s- and p-polarised species, whose respective reflectancies R_S and R_P must be computed slightly differently. To achieve an overall reflected fraction for the interface we must take the arithmetic mean of R_S and R_P.

And because we must compute reflectance at both the front and rear boundaries of the pane, for each piece of glass we have six different reflectance values to calculate: Four raw reflectancies and two averages.

The raw reflectance formulae are called the Fresnel Equations in honor of the great optician Augustin-Jean Fresnel (1788 - 1827).[5]

Incident Reflectance

Firstly, *and in terms of wave intensity (which is the square of amplitude),* and for Incidence at the front or leading boundary, and for s-polarised Reflectance, R_{Si}:-

$$R_{Si} = \left|\frac{n_1 \cdot \cos(\theta_i) - n_2 \cdot \cos(\theta_t)}{n_1 \cdot \cos(\theta_i) + n_2 \cdot \cos(\theta_t)}\right|^2$$
Equation 11

or:-

$$R_{Si} = \left| \frac{n_1 \cdot \cos(\theta_i) - n_2 \cdot \sqrt{1 - \left(\frac{n_1}{n_2} \cdot \sin(\theta_i)\right)^2}}{n_1 \cdot \cos(\theta_i) + n_2 \cdot \sqrt{1 - \left(\frac{n_1}{n_2} \cdot \sin(\theta_i)\right)^2}} \right|^2$$

Equation 12

Also, for the p-polarised Reflectance at Incidence, R_{Pi}:-

$$R_{Pi} = \left| \frac{n_1 \cdot \cos(\theta_t) - n_2 \cdot \cos(\theta_i)}{n_1 \cdot \cos(\theta_t) + n_2 \cdot \cos(\theta_i)} \right|^2$$

Equation 13

or:-

$$R_{Pi} = \left| \frac{n_1 \cdot \sqrt{1 - \left(\frac{n_1}{n_2} \cdot \sin(\theta_i)\right)^2} - n_2 \cdot \cos(\theta_i)}{n_1 \cdot \sqrt{1 - \left(\frac{n_1}{n_2} \cdot \sin(\theta_i)\right)^2} + n_2 \cdot \cos(\theta_i)} \right|^2$$

Equation 14

In practice, it is Equations Eleven and Thirteen that are computationally-convenient, and we shall use those to establish the Incident Reflectance, R_i, as the mean of R_{Si} and R_{Pi}:-

$$R_i = \frac{R_{Si} + R_{Pi}}{2}$$

Equation 15

Equations Eleven through Fifteen are all sufficiently general to operate for the important condition of normal incidence when $\theta_i = \theta_t = 0$. For that condition it is obvious that the cosine of either angle is unity, and thus:-

$$R_{Si} = R_{Pi} = \left| \frac{n_1 - n_2}{n_1 + n_2} \right|^2$$

Equation 16

whilst:-

$$R_i = \left(\frac{n_2 - n_1}{n_2 + n_1} \right)^2$$

Equation 17

Exiting Reflectance

Secondly, *and in terms of wave intensity (which is the square of amplitude),* and for Exiting at the rear or trailing boundary, and for s-polarised Reflectance, R_{Sx}:-

$$R_{Sx} = \left| \frac{n_2 \cdot \cos(\theta_t) - n_1 \cdot \cos(\theta_x)}{n_2 \cdot \cos(\theta_t) + n_1 \cdot \cos(\theta_x)} \right|^2$$

Equation 18

or:-

$$R_{Sx} = \left| \frac{n_2 \cdot \cos(\theta_t) - n_1 \cdot \sqrt{1 - \left(\frac{n_2}{n_1} \cdot \sin(\theta_t)\right)^2}}{n_2 \cdot \cos(\theta_t) + n_1 \cdot \sqrt{1 - \left(\frac{n_2}{n_1} \cdot \sin(\theta_t)\right)^2}} \right|^2$$

Equation 19

Also, for the p-polarised Reflectance at Exiting, R_{Px}:-

$$R_{Px} = \left| \frac{n_2 \cdot \cos(\theta_x) - n_1 \cdot \cos(\theta_t)}{n_2 \cdot \cos(\theta_x) + n_1 \cdot \cos(\theta_t)} \right|^2$$

Equation 20

or:-

$$R_{Px} = \left| \frac{n_2 \cdot \sqrt{1 - \left(\frac{n_2}{n_1} \cdot \sin(\theta_t)\right)^2} - n_1 \cdot \cos(\theta_t)}{n_2 \cdot \sqrt{1 - \left(\frac{n_2}{n_1} \cdot \sin(\theta_t)\right)^2} + n_1 \cdot \cos(\theta_t)} \right|^2$$

Equation 21

In practice, it is Equations Eighteen and Twenty that are computationally-convenient, and we shall use those to establish the Exiting Reflectance, R_x, as the mean of R_{Sx} and R_{Px}:-

$$R_x = \frac{R_{Sx} + R_{Px}}{2}$$

Equation 22

<u>Losses due to Absorption</u>

Absorption is the loss of light due to dissipation of the radiation in other energetic forms, especially heat. This occurs through particle excitation at the molecular level and differs from scatter, which is due to macroscopic particles.

Dimensioned Absorption, a, has the dimensions of L^{-1}, and varies with the wavelength (or conversely the frequency) of the light concerned. Visible light has a broad spectrum between about 400 and 700 nanometers but the Fraunhofer line, the Helium D_3 wavelength, is a convenient moiety at about 587.5618 nm and I have found glass data at this point.[6,7]

So for computation I have taken the Helium D_3 Wavelength to be about 0.0000005876 meters. Because both Bouguer and Buffon experimented with panes of glass each one Royal line thick I have taken my own pane thicknesses, x, to be 0.002256 meters which of course is of dimension L^{+1}. Lastly, we need a material-specific Absorption Coefficient, k, which is the Imaginary Part of the Complex Refractive Index, Ñ. The definition of Dimensioned Absorption is:-

$$a = \frac{4\pi k}{\lambda}$$

Equation 23

from which the Intensity after Absorption, I, can be computed using the exponential decay law:-

$$I = I_0 \cdot e^{-a.x}$$

Equation 24

This is Bouguer's Law. I_0 is the Input Intensity and x is Distance through which the light suffers absorption (i.e. the pane thickness).

<u>Losses due to Scatter</u>

In the case of modern glass I neglect scatter and in the case of ambient air I conflate its effects with those of absorption.

Algebraically, one may include separate scatter in the Bouguer Equation in terms of the Scatter, σ. Where scatter is not separately treated σ is zero. Note that Scatter, σ, has dimension L^{-1}.

The required modification is:-

$$I = I_0 \cdot e^{-(a+\sigma)x}$$

Equation 25

<u>Estimation of the Extinction Effect for Air</u>

By reference to literature[8] published on the Internet I assumed a conflated absorption and scatter Coefficient, b, for Poor Visibility Atmosphere of $\beta = 0.500$ km^{-1} which implies a D of eight kilometers. These data match an Atmospheric Extinction Coefficient, K_{AIR}, of 9.3519×10^{-11}.

$$b = \frac{4\pi k_{AIR}}{\lambda} \approx \frac{4\pi f k_{AIR}}{c}$$

Equation 26

The relevant Bouguer Equation is:-

$$I_T = I_0 \cdot e^{-bx}$$

Equation 27

where I_T is the Transmitted Intensity.

<u>Conflated Loss Equation</u>

By reference to the Tagirov Equations[9] I have designed a simple but accurate formula for the estimation of Total Intensity Loss to include Reflection at both the Incident and Exiting surfaces of a pane, R_i and R_x, as well as Absorption and Scatter in the body of the glass.
That Total Loss Equation is:-

$$I_T = I_0(1 - R_i)(1 - R_x)e^{-(a+\sigma)x}$$

Equation 28

This equation obviates the need to treat reflection in the air gaps (if any): The gaps require only deductions for absorption and scatter.

<u>Decay Law Identities</u>

There is considerable confusion in the literature and on the Internet regarding Bouguer's Law; its generality; its relation to Beer's Law (essentially a re-statement adapted to chemical colorimetric use); and whether or not it relates to or is a special case of the well-known Compound Interest Law of Finance.
The Compound Interest Law is:-

$$A = P\left(1 + \frac{r}{100}\right)^n$$

Equation 29

where in context A is the Accrued Amount (of currency); P is the Starting Amount (Principal); r is the Periodic Percentage Rate of Interest; and n is the Number of Lapsed Periods.

An argument is made that if the term r/100 is treated as a negative fraction (like the current interest rates at a Swiss or Japanese bank where *we* are expected to pay *them* for the privilege of using *our* money) then Equation Twenty-Nine and Bouguer's Law are the same.

In order to throw light on this matter (again no pun is intended) I constructed the spreadsheet reproduced in full in Appendix Three and entitled CASCADE ESSAYS.

Six pane-wise calculations of the surviving intensity were made in terms of $I_{T(m)}/I_0$ where $I_{T(m)}$ is the Intensity transmitted from Pane m. Table Two states the Mathematical Assumptions underlying these propositions. In all cases Original Intensity I_0 is unity:-

Column	Assumption	Implied Formula (RHS)
R	Bouguer's Implication as Stated by Buffon	$I_{T(m-1)}\left(1 - \frac{2}{7}\right)$
T	Bouguer's Law with $a = \log_n(1\text{-}2/7)$	e^{-ma}
W	Compound Interest Law	$\left(1 - \frac{2}{7}\right)^m$

Table Two
Models of Attenuation

Inspection confirms the equivalence of all three models, when the attenuation is appropriately tallied. So we may write:-

$$\frac{I_{T(m)}}{I_0} = I_{T(m-1)}\left(1 - \frac{2}{7}\right) = e^{-ma} = \left(1 - \frac{2}{7}\right)^m$$

Equation 30

or for the final pane:-

$$1 - I_5\left(1 - \frac{2}{7}\right) = 1 - \left(\frac{5}{7}\right)^6$$

Equation 31

or:-

$$I_5 \left(\frac{5}{7}\right) = \left(\frac{5}{7}\right)^6$$

Equation 32

<u>The Integrity of Buffon's Figures</u>

Column H of CASCADE ESSAYS reproduces, in decimal form, the I/I_0 series as generated from Buffon's vulgar fraction listing of Bouguer's 2/7 attenuation model. It can be seen that beyond the first pane Buffon underestimates the remaining intensities by around a percentage point, which may be an artefact of a rounding error or errors somewhere along the series.

In this context it is interesting that Buffon's Total of Fractions:-

$$\frac{102034}{117049} \cong 0.871720391$$

is different to Buffon's running total which is 0.875874016.

To study these problems in detail I developed Table Three as the worksheet BUFFON CASCADE ANOMALLY.

Buffon's statement of Bouguer's model (Concept B) may be defined as:-

$$\frac{I_m}{I_0} = \left(\frac{5}{7}\right)^m$$

Equation 33

whereas Buffon's Table One fractions' series (Concept A) generates different decrements and a different sum. The summative discrepancy may be stated as a Percentage Specific Defect, D_s, defined according to:-

$$D_s = 100 \left(\frac{\Sigma_e - \Sigma_t}{\Sigma_t}\right)$$

Equation 34

where Σ_e is the Experimental (or in this case, stated, Concept A) Sum of Data, and Σ_t the Theoretical (Concept B) Sum of Data.

Specific Defect is a signed metric of discrepancy. In this case, the quantified concepts' discrepancy is about -2%.

Also, Table Three shows that the pane-wise discrepancy A-B is strictly constant, which suggests that some systematic, theory-based controversy is at work, though Barr's Buffon is not clear on this point.

Buffon's own stated fractions for the three panes of his attempted repetition of Bouguer's experiment indicate a complete abortion, as the final I_m/I_0 is negative. But Buffon, or at least Barr, is silent about this.

Element	Serial	Denominator DR	Numerator DD	Quotient Q=DD/DR	Running Residue (1) Concept A	Buffon's Series as Running Residue (1) Value A	Buffon's Concept of Bouguer's Meaning as in Barr's Buffon Concept B	Value B	Difference of Concept Values A-B
Initial	0	1	1	1	RR0	1	$(5/7)^0$	1	0
Pane	1	7	2	0.285714286	RR0-Q1	0.714285714	$(5/7)^1$	0.714285714	0
Pane	2	47	10	0.212765957	RR1-Q2	0.501519757	$(5/7)^2$	0.510204082	-0.008684325
Pane	3	343	50	0.145772595	RR2-Q3	0.355747162	$(5/7)^3$	0.364431487	-0.008684325
Pane	4	2401	250	0.104123282	RR3-Q4	0.251623880	$(5/7)^4$	0.260308205	-0.008684325
Pane	5	16807	1250	0.074373773	RR4-Q5	0.177250107	$(5/7)^5$	0.185934432	-0.008684325
Pane	6	117649	6250	0.053124123	RR5-Q6	0.124125984	$(5/7)^6$	0.132810309	-0.008684325
Buffon Sum as stated in text		117049	102034	0.871720391					
as approximation in text		11	10	0.909090909					
Actual Sum of Fractions			102520.178	0.875874016		2.124552604		2.167974228	-0.043421624
Buffon Loss				0.871720391					
Buffon Running Residue Loss				0.875874016					
Percentage Specific Defect							$100((\Sigma A-\Sigma B)/\Sigma B)$		-2.002866243

Table Three
Buffon's Cascade Anomaly

The Effect of the Pouce Air Gap upon Buffon's Experiment

It seemed to me that Buffon's air gap was hardly worth worrying about given the vagaries of his other considerations.

But I thought the effect of that gap nevertheless a necessary item of assessment. If we assume that Buffon could have tried three contact panes then the question is one of Air gap x_2 either zero or twelve lignes, that is about 0.027072 meters.

If the Dimensioned Attenuation of dirty air is b = 0.002 then the ratio $I_{contact}/I_{pouce}$ measures the difference the air gap makes. That difference is given by:-

$$\frac{I_{contact}}{I_{pouce}} = \frac{e^{-(b+\sigma_{AIR})x_1}}{e^{-(b+\sigma_{AIR})x_2}} = e^{-(b+\sigma_{AIR}).(x_1-x_2)} = 1.00005414546581$$

Equation 35

Therefore, the pouce gap makes an indicated difference of about 0.0054% per gap.

We assume that σ_{AIR} is zero in our calculations on the grounds that σ is conflated with b, because the effects of scatter have been consolidated with those due to absorption.

The Attenuation of Light in Modern Glass and the Effect of Air Gaps

With reference to Equation Twenty-Seven for attenuation in an air gap and Equation Twenty-Eight for attenuation in a glass pane I erected the following composite model of Equation Thirty-Six:-

$$I_T = I_0 \cdot e^{-bx}$$

Equation 27

$$I_T = I_0(1 - R_i)(1 - R_x)e^{-(a+\sigma)x}$$

Equation 28

$$\frac{I_P}{I_0} = \left[(1 - R_i)(1 - R_x)e^{-(ax_1+bx_2)}\right]^P$$

Equation 36

where x_1 is the Pane Thickness and x_2 is the Gap Thickness whilst P is the Total Number of Panes. This Summary Model assumes that the Pane Thicknesses and Gap Thicknesses do not vary.

Appendix Four includes three attenuation cascade tabulations for modern Clear Soda Lime Glass (SLG) of ligne-thick panes at air gap separations of contact, one ligne and one toise; and three attenuation tabulations for Schott N-BK7 Optical Glass of ligne-thick panes at air gap separations of contact, one ligne and one toise.

Table Four summarises all six attenuation cascades whilst Figure Two is a plot of this data.

It is immediately manifest that whilst gap attenuations are finite, they are at a much reduced rate to the attenuations in glass. Furthermore, glass contact and one ligne gap attenuations are sensibly the same: Only separations of one toise (almost two meters) create a distinct effect.

The attenuation in common soda lime glass is, as any purchaser might hope, notably more rapid than in fine optical glass.

Graphical and Regression Studies

Figure Three shows the pane-wise intensity attenuation curves for the Bouguer and Buffon data, and also for modern Soda Lime and patent Optical Glass. In all cases the air gap is assumed to be zero (i.e. contact) for purposes of comparison. As aforenoted, Buffon stated that his air gaps were of one pouce each, but I have shown that makes no sensible difference to attenuation. I do not read French and accordingly have no convenient access to Bouguer's *Essai D'Optique*, but I have no reason to suspect his air gaps significant.

The catastrophic failure of Buffon's attempt to test Bouguer is manifest, including the seemingly negative attenuation of Pane Three, which I suppose implies physically that it was a glowing emitter.

Buffon's strange list of fractions supposedly owing to Bouguer's data is visibly very near to but not identical with the Bouguer line computed upon the basis of 5/7 light transmission by each pane. On the other hand the a = 2/7 model is demonstrably spurious.

The much better transmittances of the modern glasses are seen to yield almost linear attenuations within the range of computation.

Table Five presents the exponential regression data for six attenuation series plotted in Figure Three. It does of course assume the applicability of Bouguer's Law where the appropriate regression model is:-

$$y = \xi_0 e^{-\xi_1 . x}$$
Equation 37

or:-

$$\ln y = \ln \xi_0 - \xi_1 . \ln x$$
Equation 38

The regression naturally confirms that ξ_0 is unity (by definition) whilst ξ_1 identifies with the Dimensioned Absorption, a. For the modern glasses, the regression ξ_1's echo the published Internet handbook data whilst Bouguer's 2/7 loss identifies with an absorption value of a = (-)0.336472237. This differs by about a percentage point from Buffon's implication of a = 0.347733029.

<u>Half Value Layer</u>[10]

Half Value Layer (HVL) is the distance in which the transmitted light loses half of its intensity. It is defined by:-

$$HVL = \frac{\ln 0.5}{a}$$
Equation 39

The dimension of HVL is of course L^{+1}. Table Five shows that the HVL for Buffon's conception of Bouguer Glass was nearly two meters, whereas Bouguer's Model of 5/7 transmittance gives HVL = 2.060042717 meters. Modern optical Glass, in our case Schott N-BK7 has a HVL of 8.005113618 meters.

			Air	Air	Air	Air	Air	Air
Gap Material			Air	Air	Air	Air	Air	Air
Gap (meters)			0.00000000	0.00225600	1.94900000	0.00000000	0.00225600	1.94900000
Pane Material			SLG no gap	SLG ligne gap	SLG toise gap	N-BK7 no gap	N-BK7 ligne gap	N-BK7 toise gap
Pane Thickness (meters)			0.00225600	0.00225600	0.00225600	0.00225600	0.00225600	0.00225600
			Transmitted Intensity, I_T	Transmitted Intensity, I_T	Transmitted Intensity, I_T	Transmitted Intensity, I_T	Transmitted Intensity, I_T	Transmitted Intensity, I_T
		Position						
Initial		0	1.00000000	1.00000000	1.00000000	1.00000000	1.00000000	1.00000000
Pane	1	1	0.89878369	0.89878369	0.89878369	0.91705480	0.91705480	0.91705480
Gap	1	2	0.89878369	0.89877963	0.89528707	0.91705480	0.91705066	0.91348709
Pane	2	3	0.80781212	0.80780847	0.80466941	0.84098950	0.84098571	0.83771772
Gap	2	4	0.80781212	0.80780483	0.80153893	0.84098950	0.84098191	0.83445867
Pane	3	5	0.72604836	0.72604180	0.72041012	0.77123346	0.77122650	0.76524433
Gap	3	6	0.72604836	0.72603853	0.71760744	0.77123346	0.77122302	0.76226723
Pane	4	7	0.65256042	0.65255159	0.64497386	0.70726334	0.70725377	0.69904082
Gap	4	8	0.65256042	0.65254864	0.64246466	0.70726334	0.70725058	0.69632127
Pane	5	9	0.58651066	0.58650008	0.57743675	0.64859924	0.64858754	0.63856476
Gap	5	10	0.58651066	0.58649743	0.57519030	0.64859924	0.64858461	0.63608050
Pane	6	11	0.52714622	0.52713432	0.51697166	0.59480105	0.59478763	0.58332067
Gap	6	12	0.52714622	0.52713194	0.51496043	0.59480105	0.59478495	0.58105132

Table Four
Intensity Attenuation Coefficients for Modern Glass
at Air Separations of Contact, One Ligne and One Toise

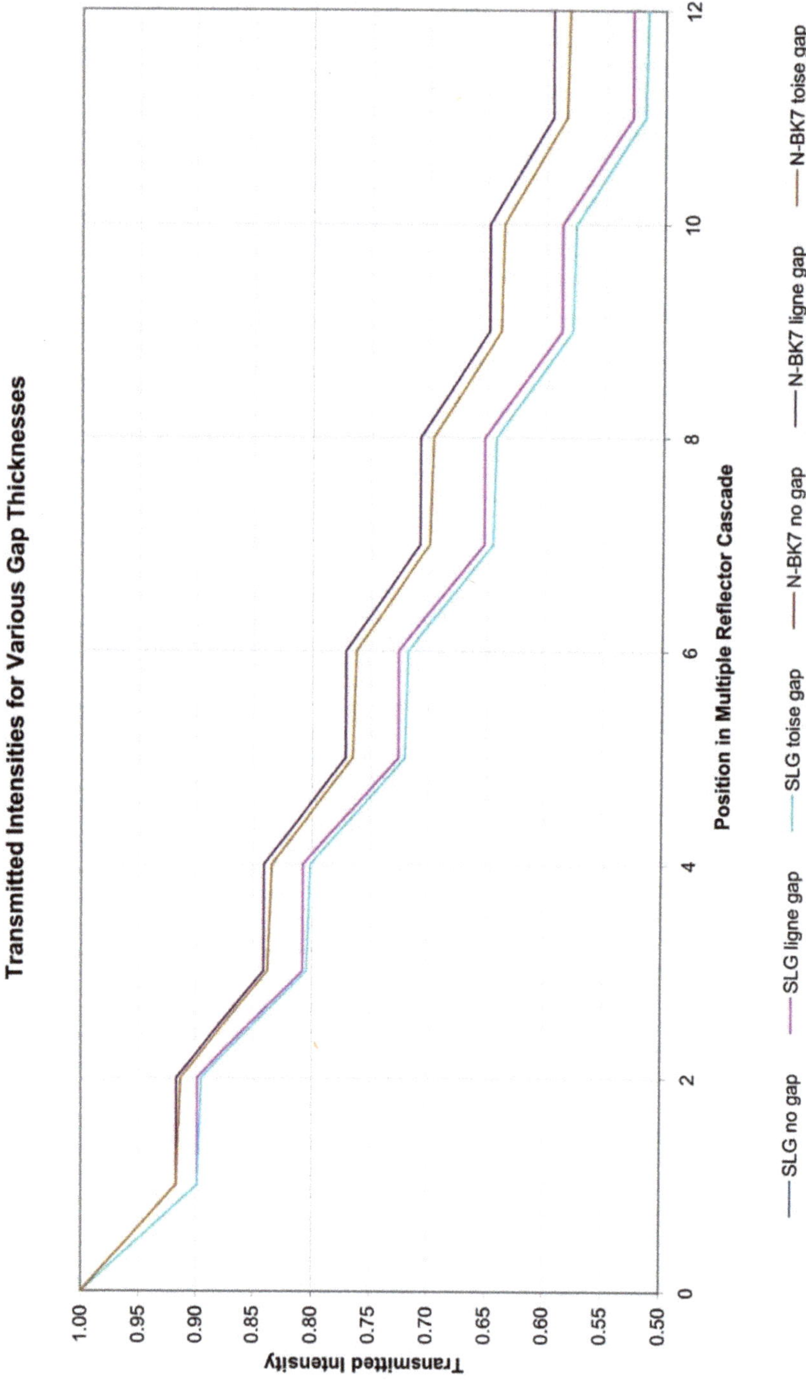

Figure Two
Intensity Attenuation Coefficients for Modern Glass
at Air Separations of Contact, One Ligne and One Toise

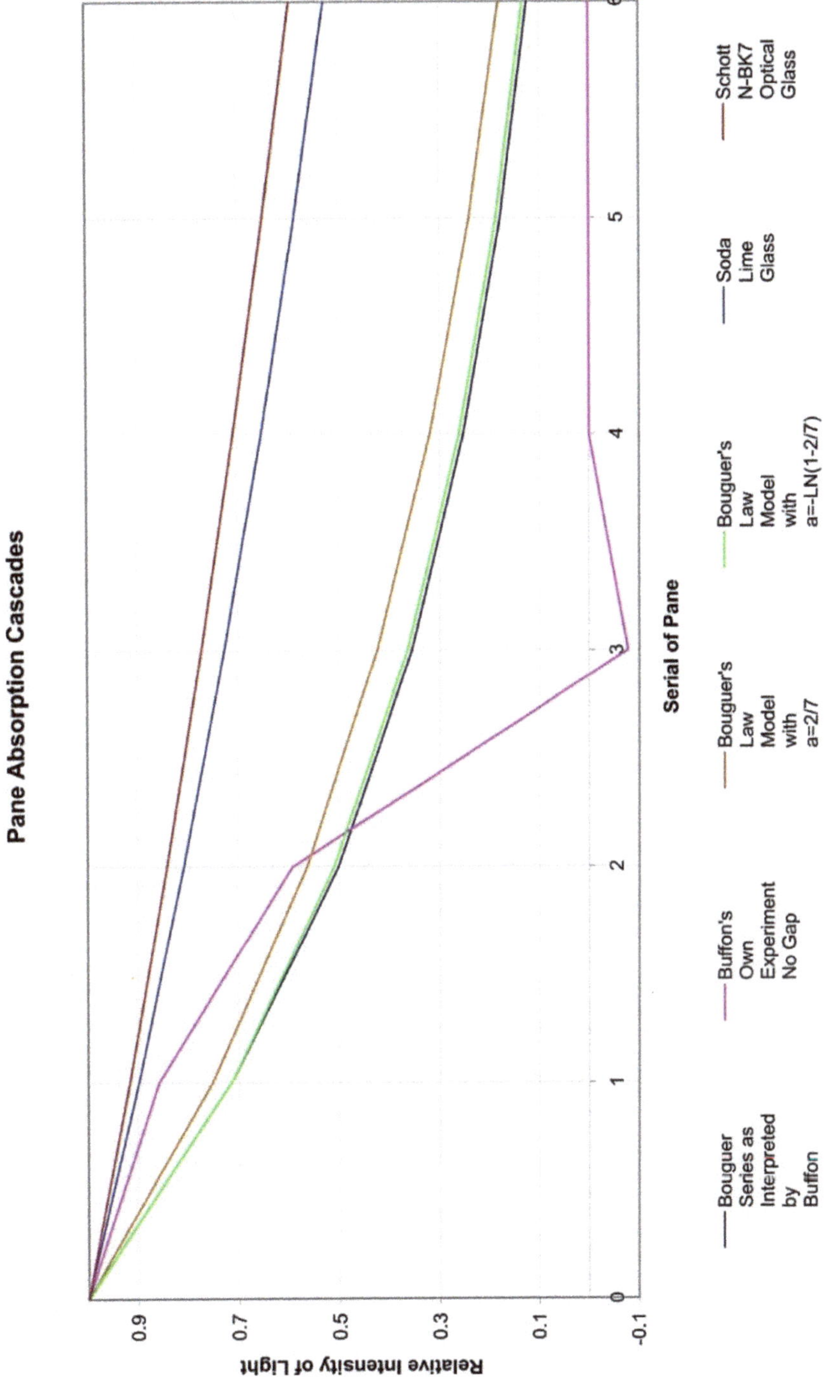

Figure Three
Intensity Attenuation for Eighteen-Century and Modern Glass
at Assumed Pane Contact

Table Five
Intensity Attenuation Exponential Regressions
for Panes in Contact

Serial	Bouguer Series as Interpreted by Buffon	Buffon's Own Experiment No Gap	Bouguer's Law Model with $a=2/7$	Bouguer's Law Model with $a=-LN(1-2/7)$	Soda Lime Glass	Schott N-BK7 Optical Glass	
0	0	0	0	0	0	0	
1	-0.336472237	-0.153025187	-0.285714286	-0.336472237	-0.106712887	-0.08658805	
2	-0.690112277	-0.522720411	-0.571428571	-0.672944473	-0.213425774	-0.173176101	
3	-1.033535019		-0.857142857	-1.00941671	-0.32013866	-0.259764151	
4	-1.379819846		-1.142857143	-1.345888946	-0.426851547	-0.346352201	
5	-1.730193508		-1.428571429	-1.682361183	-0.533564434	-0.432940251	
6	-2.08645823		-1.714285714	-2.01883342	-0.640277321	-0.519528302	
	0.006543212	0.036111673	0	2.22045E-16	-5.55112E-17	-5.55112E-17	Intercept (ln)
	-0.347733029	-0.261360206	-0.285714286	-0.336472237	-0.106712887	-0.08658805	Slope (ln) (ξ_1)
	0.999969747	0.945830968	1	1	1	1	RSQ (ln)
	0.999939494	0.894596221	1	1	1	1	RSQ^2 (ln)
	1.006564666	1.036771619	1	1	1	1	Intercept (ξ_0)
	0.70628741	0.77000351	0.751477293	0.714285714	0.898783689	0.917054798	I_1/I_0 (regression estimate)
	1.993331445	2.652076198	2.426015132	2.060042717	6.495440255	8.005113618	Half Value Layer (HVL)

References

1   "Buffon's Natural History"
(Barr's Buffon)
 Georges-Louis Leclerc
Translated from French to English by JS Barr
Volume Ten of ten volumes in English
HD Symonds of Paternoster-Row, London
1797

2   "Essai D'Optique, Sur La Gradation De La Lumiere"
Pierre Bouguer
In the Original French
Kessinger Publishing LLC   September 10 2010
ISBN 978-1166030445
pp 140
$18.36

3   "Units of measurement in France before the French Revolution"
en.wikipedia.org/wiki/
Units_of_measurement_in_France_before_the_French_Revolution

Page name: Units of measurement in France before the French
        Revolution
Author: Wikipedia contributors
Publisher: Wikipedia, The Free Encyclopedia.
Date of last revision: 8 March 2016 17:08 UTC
Date retrieved: 4 April 2016 12:14 UTC
Permanent link:
https://en.wikipedia.org/w/index.php?title=Units_of_measurement_in_France_before_the_French_Revolution&oldid=709002461
Primary contributors: Revision history statistics
Page Version ID: 709002461

4   (a)   http://www.slideplayer.com/slide/6134084/#

Chapter 2: Signals and Signal Space EENG 3810/ CSCE 3020
Instructor: Oluwayomi Adamo

(b)   "Physics: Principles with Applications"
          Seventh Edition   14 May 2015
          Douglas C Giancoli
          Chapter 11
          ISBN 978-1292057125
          Pearson Publications LLC
          pp 1080
          £50.89

5   "Fresnel equations"
          en.wikipedia.org/wiki//Fresnel_equations

          Page name: Fresnel equations
          Author: Wikipedia contributors
          Publisher: Wikipedia, The Free Encyclopedia.
          Date of last revision: 28 March 2016 06:39 UTC
          Date retrieved: 4 April 2016 14:49 UTC
          Permanent link:
https://en.wikipedia.org/w/index.php?title=Fresnel_equations&oldid=712291145
          Primary contributors: Revision history statistics
          Page Version ID: 712291145

6   Soda Lime Glass
          http://www.refractiveindex.info/
                    ?shelf=glass&book=soda-line&page=Rubin-clear

Journal citation

M. N. Polyanskiy. "Refractive index database,"

http://refractiveindex.info (accessed Feb. 30 2016).
Web link

<a href="http://refractiveindex.info"> RefractiveIndex.INFO </a>

RefractiveIndex.INFO website: © 2008-2016 Mikhail Polyanskiy
refractiveindex.info database: public domain via CC0 1.0
NO GUARANTEE OF ACCURACY - Use on your own risk

7   Schott N-BK7 Optical Glass
     http://www.refractiveindex.info/
      ?shelf=glass&book=BK7&page=SCHOTT

Journal citation

M. N. Polyanskiy. "Refractive index database,"

http://refractiveindex.info (accessed Feb. 30 2016).
Web link

<a href="http://refractiveindex.info"> RefractiveIndex.INFO </a>

RefractiveIndex.INFO website: © 2008-2016 Mikhail Polyanskiy
refractiveindex.info database: public domain via CC0 1.0
NO GUARANTEE OF ACCURACY - Use on your own risk

8   "Springer Handbook of Lasers and Optics"
    Frank Träger, editor   27 April 2007
    Hardback plus CD-ROM with Full Contents
    ISBN 978-0-387-95579-7
    pp 1332
    £145.16
    http://www.springer.com/978-0-387-95579-7

    Table 19.7; Part D19.6
    Chapter 19 Radiation and Optics in the Atmosphere
    Page 1196

9   "Lambert Formula – Bouguer Absorption Law?"
    RB Tagirov and LP Tagirov
    UDC 538:349
    Russian Physics Journal
    V40 No7 1997
    Pages 664-669

10   "Half-Value Layer"
    NDT Resource Center

    https://www.nde-ed.org/EducationResources/
    CommunityCollege/Radiography/Physics/HalfValueLayer.htm

    also:-

APPENDIX ONE

The Title Page of
Volume Ten of
Buffon's Histoire Naturelle
Translated by JS Barr

Barr's Buffon.

Buffon's Natural History.

CONTAINING

A THEORY OF THE EARTH,

A GENERAL

HISTORY OF MAN,

OF THE BRUTE CREATION, AND OF

VEGETABLES, MINERALS,

&c. &c.

FROM THE FRENCH.

WITH NOTES BY THE TRANSLATOR.

IN TEN VOLUMES.

VOL. X.

London:

PRINTED FOR THE PROPRIETOR,

AND SOLD BY H. D. SYMONDS, PATERNOSTER-ROW.

1797.

APPENDIX TWO

The Pages of Barr's Buffon
relevant to
Light Attenuation

greater inconvenience than the different refrangibility which it corrects, for these two small massive telescopes of glass, are more obscure than a small common telescope of the same glass and dimensions; they indeed give less iris, but are not better, for in massive glass the light, after having crossed this thickness of glass, would no longer have a sufficient force to take in the image of the object to our eye. So to make telescopes 10 or 20 feet long I only find water that has sufficient transparency to suffer the light to pass through this great thickness. By using, therefore, water to fill up the intervals between the objective and the ocular glass, we should in part diminish the effect of the different refrangibility, because that water approaches nearer to glass than air, and if we could, by loading the water with different salts, give it the same refringent degree of power as glass, it is not to be doubted, that we should correct still more, by this means, the different refrangibility of the rays. A transparent liquor should, therefore, be used, which would have nearly the same refrangible power as glass, for then it would be certain that the two glasses, with their liquor between them, would in part correct the effect of the different refrangibility of the rays, in the same mode as it

is

is corrected in the small massive telescope which I speak of.

According to the experiments of M. Bouguer, the thickness of a line of glass destroys $\frac{2}{7}$ of light, and consequently the diminution would be made in the following proportion:

Thickness, 1, 2, 3, 4, 5, 6 times
Diminution, $\frac{2}{7}$, $\frac{10}{49}$, $\frac{62}{343}$, $\frac{848}{2401}$, $\frac{1840}{16807}$, $\frac{6560}{117649}$.

So that by the sum of these six terms we should find, that the light which passes through six lines of glass would lose $\frac{9801}{11764}$, that is, about $\frac{12}{14}$ of its quantity. But it must be considered, that M. Bouguer makes use of glasses which are but little transparent, since he has observed, that the thickness of a line of these glasses destroys $\frac{2}{5}$ of the light. By the experiments which I have made on different kinds of white glass, it has appeared to me that the light diminishes much less. These experiments are easy to be made, and what all the world may repeat.

In a dark chamber, whose walls were blackened, and which I made use of for optical experiments, I had a candle lighted of sixes to the pound, the room was very large, and the candle the only light in it; I then tried at what distance I could read by this light, and found that I read very easily at 24 feet four inches

 from

from the candle. Afterwards, having placed a piece of glass, about a line thick, before it, at two inches distance, I found that I still read very plainly at 22 feet nine inches; and by substituting to this glass another piece of two lines in thickness, and of the same glass, I read at 21 feet distance from the candle. Two of the same glasses joined one to the other, and placed before the candle, diminished the light so much that I could only read at $17\frac{1}{2}$ feet distance; and at length, with three glasses, I could only read at 15 feet. Now the light of a candle diminishing as the square of the distance augments, its diminution should have been in the following progression, if glasses had not been interposed: 2—$24\frac{1}{3}$. 2—$22\frac{1}{4}$. 2—21. 2—$17\frac{1}{2}$. 2—15, or $592\frac{1}{2}$. $517\frac{8}{16}$ 441. $306\frac{1}{4}$. 225. Therefore the loss of the light, by the interposition of the glasses, is in the following progression: $84\frac{12}{44}$. 151. $285\frac{3}{4}$. $367\frac{1}{4}$.

From hence it may be concluded, that the thickness of a line of this glass diminishes only $\frac{12}{144}$ of light, or about $\frac{1}{7}$; that two lines diminishes $\frac{32}{77}$, not quite $\frac{1}{2}$, and three glasses of two lines, $\frac{48}{77}$, i. e. less than $\frac{2}{3}$.

As this result is very different from that of M. Bouguer, and as I was cautious of suspecting

pecting

pecting the truth of his experiments, I repeated mine with common glass. For long telescopes water can alone be used; and it is still to be feared that an inconveniency will subsist, from the opacity resulting from the quantity of liquor whic fills the interval between the two glasses.

The longer the telescope the greater loss of light will ensue; so that it appears at first sight that this mode cannot be used, especially for long telescopes; for following what M. Bouguer says in his Optical Essay, on the gradation of light, nine feet seven inches sea-water diminishes the light in a relation of 14 to 5; therefore these long telescopes, filled with water, cannot be used for observing the sun, and the stars would not have light enough to be perceived across a thickness of 20 or 30 feet of intermediate liquor.

Nevertheless, if we consider, that by allowing only an inch, or an inch and an half, for the bore of an objective of 30 feet, we shall very distinctly perceive the planets in the common telescopes of this length; we may suppose, that by allowing a greater diameter to the objective we should augment the quantity of light in the ratio of the square of this diameter, and, consequently, if an inch bore suffices to see a star distinctly, in a common telescope,

 three

APPENDIX THREE

The EXCEL® Spreadsheet Worksheet
CASCADE ESSAYS
To Test Various Purported Statements of
Bouguer's Law

	A	B	C	D	E	F	G	H
4	Pane Thickness (meters)			(Buffon's Interpretation of the Bouguer Implication)				
5				Denominator	Numerator	Quotient	Running	
6						Q	Residue (1)	
7								
8		Serial	Position					
9	Initial	0	0	1	1	1		1
10	Pane	1	1	7	2	0.285714286	RR0-Q1	0.714285714
11	Pane	2	3	47	10	0.212765957	RR1-Q2	0.501519757
12	Pane	3	5	343	50	0.145772595	RR2-Q3	0.355747162
13	Pane	4	7	2401	250	0.104123282	RR3-Q4	0.25162388
14	Pane	5	9	16807	1250	0.074373773	RR4-Q5	0.177250107
15	Pane	6	11	117649	6250	0.053124123	RR5-Q6	0.124125984
16								
17	Buffon Sum			117049	102034	0.871720391		
18				11	10	0.909090909		
19	Actual Sum of Fractions				102520.1777	0.875874016		2.124552604
20								
21	Buffon Loss					0.871720391		
22	Buffon Running Residue Loss					0.875874016		
23	Bouguer Loss					0.867189691		
24	$1-(1-2/7)^6$					0.867189691		
25	Soda Lime Loss					0.413489339		
26								
27								

	I	J	K	L	M	N
4	(Buffon's Own Experiment No Gap)					
5	Denominat	Numerator	Quotient	Running		
6			Q	Residue (2)		
7						
8						
9	1	1	1		1	
10	592	84	0.141892	RR0-Q1	0.858108	
11	592	157	0.265203	RR1-Q2	0.592905	
12	592	397	0.670608	RR2-Q3	-0.077703	
13						
14						
15						
16						
17						
18						
19						
20						
21						
22						
23						
24						
25						
26						
27						

	O	P	Q	R	S	T	U	V	W	X	Y
4										0.002256	0.002256
5	Bouguer's Law		Running		Bouguer's Law			Negative Interest		Transmitted	Transmitted
6	Model		Residue (3)		Model			Supposed Fallacy		Intensity, IT	Intensity, IT
7	With a=2/7		(Bouguer Implication)		With a=-LN(1-2/7)			With a=2/7			
8											
9	$e^{-Serial^*(2/7)}$	1		1	$e^{-Serial^*a}$	1		$(1-2/7)^{Serial}$	1	1	1
10	$e^{-Serial^*(2/7)}$	0.751477	R8(1-2/7)	0.714285714	$e^{-Serial^*a}$	0.714285714		$(1-2/7)^{Serial}$	0.714285714	0.898783689	0.917054798
11	$e^{-Serial^*(2/7)}$	0.564718	R9(1-2/7)	0.510204082	$e^{-Serial^*a}$	0.510204082		$(1-2/7)^{Serial}$	0.510204082	0.807812119	0.840989503
12	$e^{-Serial^*(2/7)}$	0.424373	R10(1-2/7)	0.364431487	$e^{-Serial^*a}$	0.364431487		$(1-2/7)^{Serial}$	0.364431487	0.726048356	0.771233459
13	$e^{-Serial^*(2/7)}$	0.318907	R11(1-2/7)	0.260308205	$e^{-Serial^*a}$	0.260308205		$(1-2/7)^{Serial}$	0.260308205	0.652560419	0.707263344
14	$e^{-Serial^*(2/7)}$	0.239651	R12(1-2/7)	0.185934432	$e^{-Serial^*a}$	0.185934432		$(1-2/7)^{Serial}$	0.185934432	0.586510661	0.648599244
15	$e^{-Serial^*(2/7)}$	0.180092	R13(1-2/7)	0.132810309	$e^{-Serial^*a}$	0.132810309		$(1-2/7)^{Serial}$	0.132810309	0.527146215	0.594801048
16											
17											
18											
19		2.479218		2.167974228		2.167974228			2.167974228		
20											
21			$(1-2/7)^6$	0.132810309							
22											
23			$1-(1-2/7)^6$	0.867189691							
24											
25											
26			2/7	0.285714286							
27			a	0.336472237							

APPENDIX FOUR

Light Attenuation Tabulations for
Modern Glass of One Ligne Pane Thickness
at Contact and at
One Ligne and One Toise Air Separations

Initial Intensity	I_0	1.00000000
Incident Light Wavelength	λ	0.00000059
Initial Angle of Incidence (degrees)	θdeg_i	0.00000000
Initial Angle of Incidence (radians)	θ_i	0.00000000
Initial Angle of Refraction (degrees)	θdeg_t	0.00000000
Initial Angle of Refraction (radians)	θ_t	0.00000000
Gap Material		Air
Gap Refractive Index	n_1	1.00010000
Gap Thickness (meters)	d_1	0.00000000
Gap Absorption Coefficient	k_1	9.3519E-11
Gap Absorption Distance	a_1	0.00199999
Pane Material		Soda Lime Glass
Pane Refractive Index	n_2	1.52340000
Pane Thickness (meters)	d_2	0.00225600
Pane Absorption Coefficient	k_2	0.00000039
Pane Absorption Distance	a_2	8.33495113

NOTES:
1 ligne = 1/144 pied du roi = 0.002256 meters
1 pied du roi = 0.03248 meters
1 toise = 6 pied du roi = 1.949 meters
"Units of Measurement in France before the French Revolution"
Wikipedia

Reflectances, R

| | | Leading Interface | | Mean | Trailing Interface | | Mean | Transmitted |
		s-polarised	p-polarised	R_i	s-polarised	p-polarised	R_x	Intensity, I_T
Initial								1.00000000
Pane	1	0.04300261	0.04300261	0.04300261	0.04300261	0.04300261	0.04300261	0.89878369
Gap	1							0.89878369
Pane	2	0.04300261	0.04300261	0.04300261	0.04300261	0.04300261	0.04300261	0.80781212
Gap	2							0.80781212
Pane	3	0.04300261	0.04300261	0.04300261	0.04300261	0.04300261	0.04300261	0.72604836
Gap	3							0.72604836
Pane	4	0.04300261	0.04300261	0.04300261	0.04300261	0.04300261	0.04300261	0.65256042
Gap	4							0.65256042
Pane	5	0.04300261	0.04300261	0.04300261	0.04300261	0.04300261	0.04300261	0.58651066
Gap	5							0.58651066
Pane	6	0.04300261	0.04300261	0.04300261	0.04300261	0.04300261	0.04300261	0.52714622
Gap	6							0.52714622

Parameter	Symbol	Value
Initial Intensity	I_0	1.00000000
Incident Light Wavelength	λ	0.00000059
Initial Angle of Incidence (degrees)	θdeg_i	0.00000000
Initial Angle of Incidence (radians)	θ_i	0.00000000
Initial Angle of Refraction (degrees)	θdeg_t	0.00000000
Initial Angle of Refraction (radians)	θ_t	0.00000000
Gap Material		Air
Gap Refractive Index	n_1	1.00010000
Gap Thickness (meters)	d_1	0.00225600
Gap Absorption Coefficient	k_1	9.3519E-11
Gap Absorption Distance	a_1	0.00199999
Pane Material		Soda Lime Glass
Pane Refractive Index	n_2	1.52340000
Pane Thickness (meters)	d_2	0.00225600
Pane Absorption Coefficient	k_2	0.00000039
Pane Absorption Distance	a_2	8.33495113

NOTES:

1 ligne = 1/144 pied du roi = 0.002256 meters

1 pied du roi = 0.03248 meters

1 toise = 6 pied du roi = 1.949 meters

"Units of Measurement in France before the French Revolution"

Wikipedia

		Reflectances, R Leading Interface		Mean	Trailing Interface		Mean	Transmitted
		s-polarised	p-polarised	R_i	s-polarised	p-polarised	R_x	Intensity, I_T
Initial								1.00000000
Pane	1	0.04300261	0.04300261	0.04300261	0.04300261	0.04300261	0.04300261	0.89878369
Gap	1							0.89877963
Pane	2	0.04300261	0.04300261	0.04300261	0.04300261	0.04300261	0.04300261	0.80780847
Gap	2							0.80780483
Pane	3	0.04300261	0.04300261	0.04300261	0.04300261	0.04300261	0.04300261	0.72604180
Gap	3							0.72603853
Pane	4	0.04300261	0.04300261	0.04300261	0.04300261	0.04300261	0.04300261	0.65255159
Gap	4							0.65254864
Pane	5	0.04300261	0.04300261	0.04300261	0.04300261	0.04300261	0.04300261	0.58650008
Gap	5							0.58649743
Pane	6	0.04300261	0.04300261	0.04300261	0.04300261	0.04300261	0.04300261	0.52713432
Gap	6							0.52713194

Parameter	Symbol	Value
Initial Intensity	I_0	1.00000000
Incident Light Wavelength	λ	0.00000059
Initial Angle of Incidence (degrees)	θdeg_i	0.00000000
Initial Angle of Incidence (radians)	θ_i	0.00000000
Initial Angle of Refraction (degrees)	θdeg_t	0.00000000
Initial Angle of Refraction (radians)	θ_t	0.00000000
Gap Material		Air
Gap Refractive Index	n_1	1.00010000
Gap Thickness (meters)	d_1	1.94900000
Gap Absorption Coefficient	k_1	9.3519E-11
Gap Absorption Distance	a_1	0.00199999
Pane Material		Soda Lime Glass
Pane Refractive Index	n_2	1.52340000
Pane Thickness (meters)	d_2	0.00225600
Pane Absorption Coefficient	k_2	0.00000039
Pane Absorption Distance	a_2	8.33495113

NOTES:

1 ligne = 1/144 pied du roi = 0.002256 meters

1 pied du roi = 0.03248 meters

1 toise = 6 pied du roi = 1.949 meters

"Units of Measurement in France before the French Revolution"

Wikipedia

		Reflectances, R Leading Interface s-polarised	p-polarised	Mean R_i	Trailing Interface s-polarised	p-polarised	Mean R_x	Transmitted Intensity, I_T
Initial								1.00000000
Pane	1	0.04300261	0.04300261	0.04300261	0.04300261	0.04300261	0.04300261	0.89878369
Gap	1							0.89528707
Pane	2	0.04300261	0.04300261	0.04300261	0.04300261	0.04300261	0.04300261	0.80466941
Gap	2							0.80153893
Pane	3	0.04300261	0.04300261	0.04300261	0.04300261	0.04300261	0.04300261	0.72041012
Gap	3							0.71760744
Pane	4	0.04300261	0.04300261	0.04300261	0.04300261	0.04300261	0.04300261	0.64497386
Gap	4							0.64246466
Pane	5	0.04300261	0.04300261	0.04300261	0.04300261	0.04300261	0.04300261	0.57743675
Gap	5							0.57519030
Pane	6	0.04300261	0.04300261	0.04300261	0.04300261	0.04300261	0.04300261	0.51697166
Gap	6							0.51496043

Parameter	Symbol	Value	
Initial Intensity	I_0	1.00000000	NOTES:
Incident Light Wavelength	λ	0.00000059	1 ligne = 1/144 pied du roi = 0.002256 meters
Initial Angle of Incidence (degrees)	θdeg_i	0.00000000	1 pied du roi = 0.03248 meters
Initial Angle of Incidence (radians)	θ_i	0.00000000	1 toise = 6 pied du roi = 1.949 meters
Initial Angle of Refraction (degrees)	θdeg_t	0.00000000	"Units of Measurement in France before the French Revolution"
Initial Angle of Refraction (radians)	θ_t	0.00000000	Wikipedia
Gap Material		Air	
Gap Refractive Index	n_1	1.00010000	
Gap Thickness (meters)	d_1	0.00000000	
Gap Absorption Coefficient	k_1	9.3519E-11	
Gap Absorption Distance	a_1	0.00199999	
Pane Material		Schott N-BK7 Optical Glass	
Pane Refractive Index	n_2	1.51680000	
Pane Thickness (meters)	d_2	0.00225600	
Pane Absorption Coefficient	k_2	0.00000001	
Pane Absorption Distance	a_2	0.20856625	

Reflectances, R

		Leading Interface		Mean	Trailing Interface		Mean	Transmitted
		s-polarised	p-polarised	R_i	s-polarised	p-polarised	R_x	Intensity, I_T
Initial								1.00000000
Pane	1	0.04214490	0.04214490	0.04214490	0.04214490	0.04214490	0.04214490	0.91705480
Gap	1							0.91705480
Pane	2	0.04214490	0.04214490	0.04214490	0.04214490	0.04214490	0.04214490	0.84098950
Gap	2							0.84098950
Pane	3	0.04214490	0.04214490	0.04214490	0.04214490	0.04214490	0.04214490	0.77123346
Gap	3							0.77123346
Pane	4	0.04214490	0.04214490	0.04214490	0.04214490	0.04214490	0.04214490	0.70726334
Gap	4							0.70726334
Pane	5	0.04214490	0.04214490	0.04214490	0.04214490	0.04214490	0.04214490	0.64859924
Gap	5							0.64859924
Pane	6	0.04214490	0.04214490	0.04214490	0.04214490	0.04214490	0.04214490	0.59480105
Gap	6							0.59480105

Parameter	Symbol	Value
Initial Intensity	I_0	1.00000000
Incident Light Wavelength	λ	0.00000059
Initial Angle of Incidence (degrees)	θdeg_i	0.00000000
Initial Angle of Incidence (radians)	θ_i	0.00000000
Initial Angle of Refraction (degrees)	θdeg_t	0.00000000
Initial Angle of Refraction (radians)	θ_t	0.00000000
Gap Material		Air
Gap Refractive Index	n_1	1.00010000
Gap Thickness (meters)	d_1	0.00225600
Gap Absorption Coefficient	k_1	9.3519E-11
Gap Absorption Distance	a_1	0.00199999
Pane Material		Schott N-BK7 Optical Glass
Pane Refractive Index	n_2	1.51680000
Pane Thickness (meters)	d_2	0.00225600
Pane Absorption Coefficient	k_2	0.00000001
Pane Absorption Distance	a_2	0.20856625

NOTES:

1 ligne = 1/144 pied du roi = 0.002256 meters

1 pied du roi = 0.03248 meters

1 toise = 6 pied du roi = 1.949 meters

"Units of Measurement in France before the French Revolution"

Wikipedia

Reflectances, R

		Leading Interface		Mean	Trailing Interface		Mean	Transmitted
		s-polarised	p-polarised	R_i	s-polarised	p-polarised	R_x	Intensity, I_T
Initial								1.00000000
Pane	1	0.04214490	0.04214490	0.04214490	0.04214490	0.04214490	0.04214490	0.91705480
Gap	1							0.91705066
Pane	2	0.04214490	0.04214490	0.04214490	0.04214490	0.04214490	0.04214490	0.84098571
Gap	2							0.84098191
Pane	3	0.04214490	0.04214490	0.04214490	0.04214490	0.04214490	0.04214490	0.77122650
Gap	3							0.77122302
Pane	4	0.04214490	0.04214490	0.04214490	0.04214490	0.04214490	0.04214490	0.70725377
Gap	4							0.70725058
Pane	5	0.04214490	0.04214490	0.04214490	0.04214490	0.04214490	0.04214490	0.64858754
Gap	5							0.64858461
Pane	6	0.04214490	0.04214490	0.04214490	0.04214490	0.04214490	0.04214490	0.59478763
Gap	6							0.59478495

Initial Intensity	I_0	1.00000000
Incident Light Wavelength	λ	0.00000059
Initial Angle of Incidence (degrees)	θdeg_i	0.00000000
Initial Angle of Incidence (radians)	θ_i	0.00000000
Initial Angle of Refraction (degrees)	θdeg_t	0.00000000
Initial Angle of Refraction (radians)	θ_t	0.00000000
Gap Material		Air
Gap Refractive Index	n_1	1.00010000
Gap Thickness (meters)	d_1	1.94900000
Gap Absorption Coefficient	k_1	9.3519E-11
Gap Absorption Distance	a_1	0.00199999
Pane Material		Schott N-BK7 Optical Glass
Pane Refractive Index	n_2	1.51680000
Pane Thickness (meters)	d_2	0.00225600
Pane Absorption Coefficient	k_2	0.00000001
Pane Absorption Distance	a_2	0.20856625

NOTES:

1 ligne = 1/144 pied du roi = 0.002256 meters

1 pied du roi = 0.03248 meters

1 toise = 6 pied du roi = 1.949 meters

"Units of Measurement in France before the French Revolution"

Wikipedia

| | | Reflectances, R | | | | | | Transmitted |
| | | Leading Interface | | Mean | Trailing Interface | | Mean | |
		s-polarised	p-polarised	R_i	s-polarised	p-polarised	R_x	Intensity, I_T
Initial								1.00000000
Pane	1	0.04214490	0.04214490	0.04214490	0.04214490	0.04214490	0.04214490	0.91705480
Gap	1							0.91348709
Pane	2	0.04214490	0.04214490	0.04214490	0.04214490	0.04214490	0.04214490	0.83771772
Gap	2							0.83445867
Pane	3	0.04214490	0.04214490	0.04214490	0.04214490	0.04214490	0.04214490	0.76524433
Gap	3							0.76226723
Pane	4	0.04214490	0.04214490	0.04214490	0.04214490	0.04214490	0.04214490	0.69904082
Gap	4							0.69632127
Pane	5	0.04214490	0.04214490	0.04214490	0.04214490	0.04214490	0.04214490	0.63856476
Gap	5							0.63608050
Pane	6	0.04214490	0.04214490	0.04214490	0.04214490	0.04214490	0.04214490	0.58332067
Gap	6							0.58105132

Barr's Buffon
and the
The Globes of Monsieur Tillet

by

James R Warren BSc MSc PhD PGCE

Sometime in the middle of the eighteenth century, in France, Georges-Louis Leclerc, First Comte de Buffon consulted Monsieur Tillet of The Academy of Sciences.

After some discussion of his opinion that what we should now call specific heat bore no relation to material mass density, Buffon digressed to describe how he caused Tillet to order the manufacture of one-inch diameter spheres of sundry metallic and other materials, which spheres Tillet weighed[1].

Why the two savants did this is not clear to me, especially since the making of a globe by any method is troublesome, precision difficult to attain, and several high skills and extended cares necessary. It crossed my mind that the entire experiment may have had a ballistic motivation, and state subvention, but Buffon's papers are of course silent of any such intent.

Buffon was, however, prompt to admit that no-one should use Tillet's globes in the calculation of material mass densities, and his reasons are manifest in the nevertheless remarkably interesting table of globe weights he supplies[2].

For example, the fifteenth entry in that table is for "Calcareous white ftone of the quarry of Anieres, near Dijon" which allegedly weighed six onces, six denier scruples and six grains. By comparison, the ball of gold weighed 6oz, 2d and 17 grains!

And this is not all: "Rock chryftal; it was a little too fmall, and had many defects. I prefume that without them it would have weighed – 0 6 22".

Notwithstanding these amusing observations I mean no mockery, because Tillet's table bears some deep implications and evokes some fascinating thoughts regarding manufacture, precision and the reliability of data.

<u>Ancient and Modern French Measures[3]</u>

The French Revolution took place in 1789. Over the next ten years the French replaced their customary weights and measures with a decimal system based upon the length of a terrestrial meridian, and the weight of a cubic meter (one meter was declared to be 1/40000000 of a meridian) of pure water. At the time these were assumed invariant.

The old system was very similar in unit sizes to the English system of measures, but even then more rational. Both old systems were based upon Roman standards.

The new system, with some extensions and minor adjustments, forms the *Système international d'unités*, SI, now universally used.

Table One presents the mid-eighteenth century Parisian French weight (mass) measures as we might assume used by Tillet, Buffon and other gentleman savants of that place and time.

Unit	Numerator	Divisor	Quotient	Calculated Value (kgs)	Wikipedia Value (kgs)
Livre	1	1	1		0.4895
Once	1	16	0.0625	0.03059375	0.03059
Denier Scruple	1	384	0.002604167	0.00127474	0.001275
Grain	1	9216	0.000108507	0.00005311	0.00005311

Table One
Ancien Regime Mass Measures

Table Two gives the three key contemporary measures of length. The Pied de Roi and Pouce were respectively very similar to the (British) Imperial foot and inch used as late as 1980 in the UK, and still used in modified forms in the US.

Unit	Numerator	Divisor	Quotient	Calculated Value (m)	Wikipedia Value (m)
Pied du Roi	1	1	1		0.3248
Pouce	1	12	0.083333333	0.027066667	0.02707
Ligne	1	144	0.006944444	0.002255556	0.002256

Table One
Ancien Regime Length Measures

<u>Globe Volumes</u>

In spite of Buffon's cautions we shall assume that each specimen was a geometrically-perfect sphere of one Parisian pouce diameter.

The Parisian pouce is 0.027066667 meters.

Hence, the globe volumes are:-

$$V = \frac{4}{3}\pi \left(\frac{D}{2}\right)^3 = 0.0000103825244 \; m^3$$

Equation 1

where V = Globe Volume (m^3); π is The Ludolphine Constant; and D is the Globe Diameter (one Parisian Pouce).

<u>Mass Density</u>

Notwithstanding Buffon's warning, and our own natural misgivings, it would be remiss not to compute the Indicated Densities, ρ_i. of the Tillet Globes, and compare these with the Fiducial Densities, ρ_f, from modern references.

In particular, the simple statistic Percentage Specific Defect, %η, may be defined as:-

$$\%\eta = 100\left(\frac{\rho_i - \rho_f}{\rho_f}\right)$$

Equation 2

Table Three presents the original globe data of Page 126 in Barr's Buffon Volume Ten, together with the Computed Tillet Density, ρ_i; the Reference Density, ρ_f, and the Specific Defect, %η.

Tin allotrope β (white tin) is assumed, and the Reference Density of Antimony does of course apply to the material without vesicles.

With regard to Specific Defects two broad facts are salient:-

(a)   Tillet Densities underestimate Fiducial Densities except for
        White Marble
        (%η's are negative)

(b)   Whilst the specific defects of metals average -14% those of
        non-metallic substances average -42%

In view of the clear differences we will define two distinct materials' groups:-

(I)   Group A:- Metals
        Gold, Lead, Silver, Bismuth, Copper, Iron, Tin, Antimony

(II)   Group B:- Non-Metals
        Emery, Marble, Clay, Gypsum, Rock Crystal, Glass,
        Earth, Ocher, Porcelain, Chalk, Cherrywood

I have omitted certain irrecoverable enigmas such as "Fine", "Marble common of Montbard" and the outrageous "Calcareous white stone of the quarry of Anieres, near Dijon".

It is notable that the last, whatever it was, was mined from Buffon's own estate, suggestive of a very close collaboration between Buffon and Tillet.

Until the latter years of the nineteenth-century, when international copyright regulations began to be workable, plagiarism was endemic within the worlds of printing and publishing. In order to assist the forensic proof of priority printers would sometimes interpolate stray characters, misspellings or nonsense in their products, even if scientific disquisitions. If the weight of the "Calcareous white stone" ball is such it is not particularly subtle as 6oz. 6d. 6g.: The "Number of the Beast" from the Book of Revelation[4].

But such tricks were not intended to confuse scientists or theologians, often the same thing in those days.

Figure One is a graphical illustration of the relatively close agreement of the metallic Tillet densities with modern values, as contrasted with the erratic dispersion of non-metals. Percentage Specific Defect is plotted along the ordinate, with the zero line of exact agreement. Note that the abscissa is wholly arbitrary, furnishing only placeholders for the various substances.

<u>Aspects of Precision</u>

There are an infinite number of ways in which precision may be quantified. But the concept is fraught with problems and some of precision's possible mathematical formulations may yield counter-intuitive, or indeed useless, results.

For example, we may declare Precision, Ω, to be a joint function of both Manufacturing Technique, M, and Material Competency, C:-

$$\Omega = f(M).g(C)$$
Equation 3

Manufacturing Techniques for Spheres

It is much more difficult to craft a sphere than a cuboid, which presents another question as to the motives and conduct of Tillet's experiment.

Methods accessible to pre-industrial Europeans include these broad techniques, not suited to every material:-

    (a)   Cutting

Suited to competent, elastic substances like metal, wood or some limestones the classic method is to turn a cylindrical billet in a lathe whilst rotating the cutter around a center below the billet's medial line. This action, carefully and gradually orbited, produces a geometrically-perfect sphere attached by a residual stalk which may gingerly be detached to finish.

    (b)   Moulding

A plastic round rod, for example white-hot steel feedstock can be fed through skew rollers which break the rod into uniform spheres that may be stabilised in a quench bath.

I doubt that this method, suited to the mass production of bearing balls, would have been practical for eighteenth-century laboratories

(c) Forging

A cold or hot round rod is cut into uniform cylinders of computed length. These smaller roundels are then struck to shape in the hemispherical jaws of an adapted power hammer.

(d) Casting

Molten material is poured into a void within a shaped refractory mould. The mould is pre-shaped as a sphere or two mated hemispheres. When the cast has cooled it is freed from the mould and the sprue and flashing ground away.

Not suited to any of Tillet's non-metal specimens, and problematic for antimony and bismuth.

(e) Surface Effect Forming

This involves the insertion of a moving liquid material into a second, immiscible fluid.

The classic eighteenth-century application of this manufacturing technique was the forming of spherical lead shot by dropping molten lead from a sieve. The lead droplets fell within a high (air-filled) tower into a cooling and decelarative water bath.

Suited to some plastic, easily-fused metals.

(f) Faceting

This involves the grinding of a cube into polyhedra that successively approximate a sphere.

A very crude method, utterly dependent upon craft skill, but a method inferred though not stated in Barr's Buffon.

This lapidary method could equally be applied to metals and non-metals, but spoilings and shatterings are very probable.

This is not an exhaustive list of possible sphere-forming methods, even for the eighteenth century.

The list is arranged roughly in descending order of control and precision.

15:03 Thursday, 29 September 2022
James R Warren

Unit Mass (kgs)		0.03	0	0					
Material	oz.	d.	gns.		Globe Mass (kilograms)	Computed Tillet Density (kg/m^3)	Reference Density (kg/m^3)	Percentage Specific Defect	Notes	Group
Gold	6	2	17	0.18701492	18012.47161	19320	-6.767745303		A
Lead	3	6	28	0.10091688	9719.879595	11389	-14.65554838		A
Pure Silver	3	3	22	0.09677398	9320.852959	10500	-11.22997182		A
Bismuth	3	0	3	0.09194059	8855.321884	9747	-9.148231418		A
Copper-red	2	7	56	0.07308507	7039.239117	8960	-21.43706342		A
Iron	2	5	10	0.06809234	6558.360864	7874	-16.70865045		A
Tin	2	3	48	0.06756120	6507.203603	7310	-10.98216686	(α) 5750 (β) 7310	A
Antimony (cavitated)	2	1	34	0.06426812	6190.028584	6691	-7.487242801		A
Emery	1	2	24.5	0.03444453	3317.548378	3650	-9.108263628	corundum	B
White Marble	1	0	25	0.03192160	3074.551388	2550	20.57064265		B
Pure Clay	0	7	24	0.01019792	982.2194117	1089	-9.805380009		B
White Gypsum	0	6	36	0.00956055	920.8306985	2787	-66.95978836		B
Rock Crystal (defective)	0	6	22	0.00881695	849.210533	1390	-38.90571705	flint	B
Common Glass	0	6	21	0.00876383	844.0948069	2579	-67.27046115		B
Pure Dry Earth	0	6	16	0.00849826	818.5161764	1249	-34.46627891	dry loam	B
Ocher	0	5	9	0.00685173	659.9286672	2700	-75.55819751	limonite: sg=2.7-4.3	B
Lauraguais Porcelain	0	5	2.5	0.00650648	626.6764476	2403	-73.92108		B
White Chalk	0	4	49	0.00770155	741.7802849	2499	-70.31691537		B
Cherrywood	0	1	59	0.00440847	424.6052665	630	-32.60233865		B

Mean Percentage Specfic Defect (All)		-29.30317886
Mean Percentage Specfic Defect (Metals)		-12.30207756
Mean Percentage Specfic Defect (Non-Metals)		-41.66761618

Table Three
Mass Density Differentials for
Tillet's Globe Materials

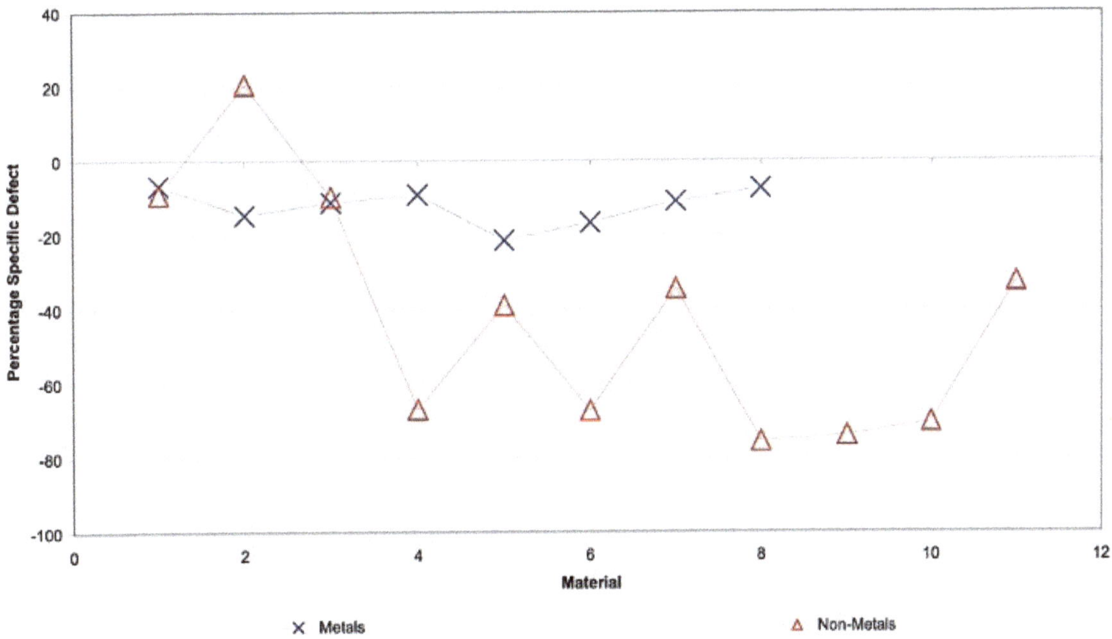

Figure One
Tillet's Globes Mass Density Defects

The Competency of Materials

"Competence" or "Competency" is an old and suitably vague geological term.

It means the strength of coherence of a material, in the sense of resistance to erosion.

It may, and often does, involve such quantifiable properties as shear, compression and tensile strengths; size, fabric and habit of crystalline constituents; plasticity; and ability to become liquid, whether by stress or dissolution.

In the Tillet context the metals exhibit malleability and ductility, as well as low melting point and chemical stability under fusion and solidification, which together promote precision machining. So besides mechanical strength characteristics, it may be relevant to consider thermal properties like specific heat, thermal conductivity, and melting point.

On the other hand, the non-metals exhibit a friability and refractoriness that make cohesion problematic, especially under stress. I would genuinely be interested to know how one fashions a one-inch sphere of "Pure earth, very dry" or even "White chalk".

As a first essay at the statistical quantification of Precision I computed the Differentials, δ, of the two series Tillet Density, ρ_i, and corresponding Fiducial Density, ρ_f, defined as:-

$$\delta_j = \rho_{f,j} - \rho_{i,j}$$
Equation 4

where j is a Serial Subscript denoting a specific material.

From a statistical point of view the Series of Tillet Densities $\rho_{i,j}$ constituted FALSE Group 1; and the Series of Fiducial Densities $\rho_{f,j}$ constituted TRUE Group 2.

N_1 was the Number of Group 1 Members and N_2 the Number of Group 2 Members. Where $N = N_1 = N_2$ there were matched series pairs and other statistical metrics were calculable.

Specifically, the Variation, σ_{rel}, of the Differences is given by:-

$$\sigma_{rel} = \frac{\sigma_\delta}{|\mu_\delta|} = \sqrt{\frac{\sum_{j=1}^{N}(\delta_j - \mu_j)^2}{N}} \times \frac{1}{\left|\dfrac{\sum_{j=1}^{N}\delta_j}{N}\right|} = \sqrt{\pi} \cdot \frac{\sqrt{\sum_{j=1}^{N}(\delta_j - \mu_j)^2}}{\sum_{j=1}^{N}\delta_j}$$
Equation 5

where the Mean Difference, μ_j, is given by:-

$$\mu_j = \frac{\sum_{j=1}^{N}\delta_j}{N}$$
Equation 6

and the Population Standard Deviation of the Differences, σ_δ, is given by:-

$$\sigma_\delta = \sqrt{\frac{\sum_{j=1}^{N}(\delta_j - \mu_\delta)^2}{N}}$$
Equation 7

Unless the data is paired (i.e. $N_1 = N_2$), the statistics of Equations Five, Six and Seven are undefined.

Using Variation it is possible to define a strictly statistical metric of Precision, Ω, as:-

$$\Omega = \frac{1}{1 + \sigma_{rel}}$$
Equation 8

Equation Eight has the virtue that as σ_{rel} Variation becomes small, Ω converges to unity for absolute exactitude of precision, whilst as variation becomes large Precision tends to, but never becomes, zero.

Table Four summarises the precision statistics for Tillet's globes. It is clear that in these terms the metallic globes are much more precisely machined, or rather more realistically-dense, than the non-metallic.

	Metals	Non-Metals
Variability, σ_{rel}	0.36207838	0.92701333
Precision, Ω	0.73417214	0.51893777

Table Four
The Relative Variations and Precisions of
Tillet's Metal and Non-Metal Globes in
Terms of Mass Density Differences

<u>The Computation of the Gosset ("Student's") t-Statistic</u>

We may now pose that notorious question unintelligible to the vast majority of intelligent laymen:-

"The Tillet densities are FALSE and the corresponding modern densities, whilst admittedly inaccurate, are TRUE, but are the two sets of figures *different*?"

The Gosset or "Student" probability distribution is a modified form of the Gaussian (or "Normal") probability distribution suited to comparing small samples of not necessarily equal numbers of items.

The classical Gosset treatment assumes that whilst Sample Means μ_1 and μ_2 differ the Sample Variances s_1 and s_2 are the same, i.e. $s_1 = s_2$. This latter criterion is usually unrealistic, and indeed we know that the variances of the FALSE Tillet densities and the TRUE fiducial densities differ.

The Gosset family of statistical distributions relate three (dimensionless) statistical parameters:-

(a)   Degrees of Freedom, ν

This defines the (symmetrical) shape of the sample distribution. As ν approaches infinity (or in practice a very large sample, perhaps as small as fifty items): The Gosset Distribution approaches the Gaussian Distribution. For our purposes the Degrees of Freedom is one less than the Number of Data, N:-

$$\nu = N - 1$$
Equation 9

(b)   Fractional Threshold Statistic, t

This relates a Confidence Limit against which a computed Φ probability may be compared to accept or reject a formal hypothesis that compared samples are abstracts of a common population (i.e. they are "the same").

Meaningful t is between a very small fraction (say 10^{-6}) and unity.

(c)   Distributional Probability $\Phi(t|\nu)$

As implied above, Φ is a function of ν and t. The Probability Φ can be used to express the likelihood (on a scale of 0 to 1) that (two) compared samples satisfy a hypothesis, which is typically that they do not "differ", because they are extracts of the same inclusive population.

Our intent is slightly different in spirit. We wish to compute the fractional agreement between each of the Tillet density lists and their modern density values.

This will give us a second metric of the precision of globe fabrication.

To do this we will apply a so-called "two-tailed test".

As aforementioned, the classic Gosset Test assumes equal sample variances, which we know to be wrong. Therefore we will apply the Welch Test[5], a variant of the Gosset Test based upon the Welch-Satterthwaite Equation, and which allows for unequal sample variances.

Some of the mathematical aspects[6] of the Gosset and Welch Tests are reviewed in Appendix Two.

Conveniently, we do not need to select either ν or t for the Welch Test, as the data mathematics establishes them for us, and indeed Welch ν can be any positive real, not necessarily an integer.

Please note that I have not tested whether either the Tillet or modern density distribution are Gaussian: Both Gosset and Welch assume all data to be Gaussian.

Welch t

First, t is computed according to Welch practice:-

$$t = \left| \frac{X_1 - X_2}{\sqrt{\dfrac{s_1{}^2}{N_1} + \dfrac{s_2{}^2}{N_2}}} \right|$$

Equation 10

where X_1 and X_2 are the First and Second Sample Means; $s_1{}^2$ and $s_2{}^2$ are the respective Sample Variances; and N_1 and N_2 are the First and Second Sample Sizes.

For our paired densities the sample sizes are of course the same.

Welch ν

This is technically an approximation, but a deadly accurate one:-

$$\nu \approx \frac{\left(\dfrac{s_1^{\,2}}{N_1} + \dfrac{s_2^{\,2}}{N_2}\right)^2}{\dfrac{\left[\dfrac{s_1^{\,2}}{N_1}\right]^2}{\nu_1} + \dfrac{\left[\dfrac{s_2^{\,2}}{N_2}\right]^2}{\nu_2}}$$

Equation 11

Clearly, Equation Eleven is susceptible of computational simplification, including in light of the fact that $\nu_1 = \nu_2$ for our paired data.

Whilst ν_1 and ν_2 are the same integer, it is manifest that ν can be, and in practice always is, a positive real number with a fractional addend.

Welch $\Phi(t|\nu)$

The Distributional Probability is computed by reference to a definite integral in t, which has no analytic solution. It is scaled by a factor involving Gamma Functions of the Equation Eleven degrees of freedom.

Mathematical details are discussed in Appendix Two.

For comparing our densities' series we will employ the EXCEL 2000® intrinsic function for Gosset Probability TDIST, which has the form:-

=TDIST(t,ν,tails)

which of course we enter as:-

=TDIST([t cell],[ν cell],2)

Modern spreadsheets offer slightly-variant functional syntax.

Welch Φ as a Metric of Association

The design purpose of the Welch Test, like that of any statistical hypothesis test is to establish, within confidence limits expressed by t, whether or not two samples derive from a common population.

Our purpose is to express, in fractional or percentage terms, the similarity of a suspect collection of data to a corresponding set of fiducial values.

Table Five presents sample statistics, precisions, and Welch Test results for Tillet and Fiducial metal mass densities. Table Six presents the same data and statistics for Tillet and Fiducial non-metallic mass densities.

The Welch Probability $p = A(t|\nu) = \Phi(t|\nu)$ for metals is 0.554277602311, suggesting a 55% similarity of Tillet to Fiducial mass densities.

The corresponding $\Phi(t|\nu)$ for non-metals is 0.033599445992 which is a 3% degree of similarity.

Of course, except for validation purposes, we are unable to take the full twelve-figure result too seriously: The "true" modern values of mass density are known to be provisional, often only recorded to two-figure accuracy, and of course very sensitive to variations in temperature, chemistry and crystallography.

WELCH'S UNEQUAL VARIANCES t-TEST
PHASE ONE: GROUP A COMPARISON (METALS)

TAILS (1 or 2): 2

Computed Values:

t	0.607042691042	
ν (approx)	13.986765575508	Welch-Satterthwaite Equation
p=A(t|ν) (TDIST)	0.554277502311	
t (TINV)	0.607042238698	

Mean of Differences, μ_δ	1198.455223416860	
Variability, σ_{rel}	0.362078380257	$\sigma_{rel} = (\Sigma(\delta-\mu_\delta)^2/N_2)^{0.5}/\|\mu\|$
Precision, Ω	0.734172140528	$\Omega = 1/(1+\sigma_{rel})$

Group 1 Tillet's Series
Group 2 Fiducial Series

	Group 1 FALSE	Group 2 TRUE	
μ_1	9025.42	10223.88	μ_2
σ_1	3636.237	3749.864	σ_2
N_1	8	8	N_2
ν_1	7	7	ν_2
$Xbar_1$	9025.42	10223.88	$Xbar_2$
s_1^2	15111105	16070261	s_2^2
s_1^2/N_1	1888888	2008783	s_2^2/N_2

	Serial	Group 1 FALSE	Group 2 TRUE	Difference, δ	$(\delta-\mu_\delta)^2$
Gold	0	18012.47	19320	1307.52839	1709630
Lead	1	9719.88	11389	1669.1204	2785963
Pure Silver	2	9320.853	10500	1179.14704	1390388
Bismuth	3	8855.322	9747	891.678116	795089.9
Copper-red	4	7039.239	8960	1920.76088	3689322
Iron	5	6558.361	7874	1315.63914	1730906
Tin	6	6507.204	7310	802.796397	644482.1
Antimony (cavitated)	7	6190.029	6691	500.971416	250972.4

Table Five
Phase One: Group A Comparison (Metals)
Sample Statistics, Precisions and Welch Test Results

WELCH'S UNEQUAL VARIANCES t-TEST
PHASE TWO: GROUP B COMPARISON (NON-METALS)

TAILS (1 or 2):   2

Computed Values:

t	2.290506701122	
ν (approx)	19.840465028280	Welch-Satterthwaite Equation
p=A(t|v) (TDIST)	0.033599445992	
t (TINV)	2.290507836733	

Mean of Differences, μ_δ	933.276176541486	
Variability, σ_{rel}	0.927013327381	$\sigma_{rel} = (\Sigma(\delta-\mu_\delta)^2/N_2)^{0.5}/|\mu|$
Precision, Ω	0.518937770586	$\Omega = 1/(1+\sigma_{rel})$

Group 1   Tillet's Series
Group 2   Fiducial Series

	Group 1 FALSE	Group 2 TRUE	
μ_1	1205.451	2138.727	μ_2
σ_1	951.0674	869.2862	σ_2
N_1	11	11	N_2
ν_1	10	10	ν_2
$Xbar_1$	1205.451	2138.727	$Xbar_2$
s_1^2	994982.2	831224.4	s_2^2
s_1^2/N_1	90452.93	75565.86	s_2^2/N_2

	Serial	Group 1 FALSE	Group 2 TRUE	Difference, δ	$(\delta-\mu_\delta)^2$
Emery	0	3317.548	3650	332.451622	110524.1
White Marble	1	3074.551	2550	-524.551388	275154.2
Pure Clay	2	982.2194	1089	106.780588	11402.09
White Gypsum	3	920.8307	2787	1866.1693	3482588
Rock Crystal (defective)	4	849.2105	1390	540.789467	292453.2
Common Glass	5	844.0948	2579	1734.90519	3009896
Pure Dry Earth	6	818.5162	1249	430.483824	185316.3
Ocher	7	659.9287	2700	2040.07133	4161891
Lauraguais Porcelain	8	626.6764	2403	1776.32355	3155325
White Chalk	9	741.7803	2499	1757.21972	3087821
Cherrywood	10	424.6053	630	205.394733	42187

Table Six
Phase Two: Group B Comparison (Non-Metals)
Sample Statistics, Precisions and Welch Test Results

<u>The Visualisation of Tillet's Globes</u>

After an Indicated Diameter, D, has been computed for each of Tillet's globes it is possible to use the EXCEL 2000® graphical tool:-

INSERT>CHART>BUBBLE

to create a bubble chart sheet.

Radius Computation

The EXCEL bubble chart diameters are of course twice the globe radius:-

$$D = 2r_t$$

Equation 12

where r_t is the Tillet radius of the ith globe. r_t is given by:-

$$r_t = \sqrt[3]{\frac{3}{4} \cdot \frac{V}{\pi}} = \sqrt[3]{\frac{3}{4} \cdot \frac{M}{\rho_f \cdot \pi}}$$

Equation 13

where V and M are respectively the Volume and Mass of the ith globe and ρ_f is the modern Fiducial Mass Density of the relevant globe. π is of course the Ludolphine Constant, about 3.14159265, a figure conveniently accessible via the Weierstrass Integral:-

$$\pi = \int_{-1}^{+1} \frac{dx}{\sqrt{1 - x^2}}$$

Equation 14

Presentations

The size of Tillet's globes is visually represented in Figure Two in which the bubble width is directly proportional to Tillet Diameter, D.

The metal series Group A is represented by blue bubbles and non-metallic substances by red.

Please note that the horizontal axis is purely ordinal and that in particular no inference of linearity can be made regarding the metal dispositions.

The relative consistency of the metal products, for whatever reasons, is readily apparent.

Appendix Four tabulates globe size data which show, in terms of pouce, that metal globes are 2.3-8.3% too big, whilst non-metallic globes are around 50% too large.

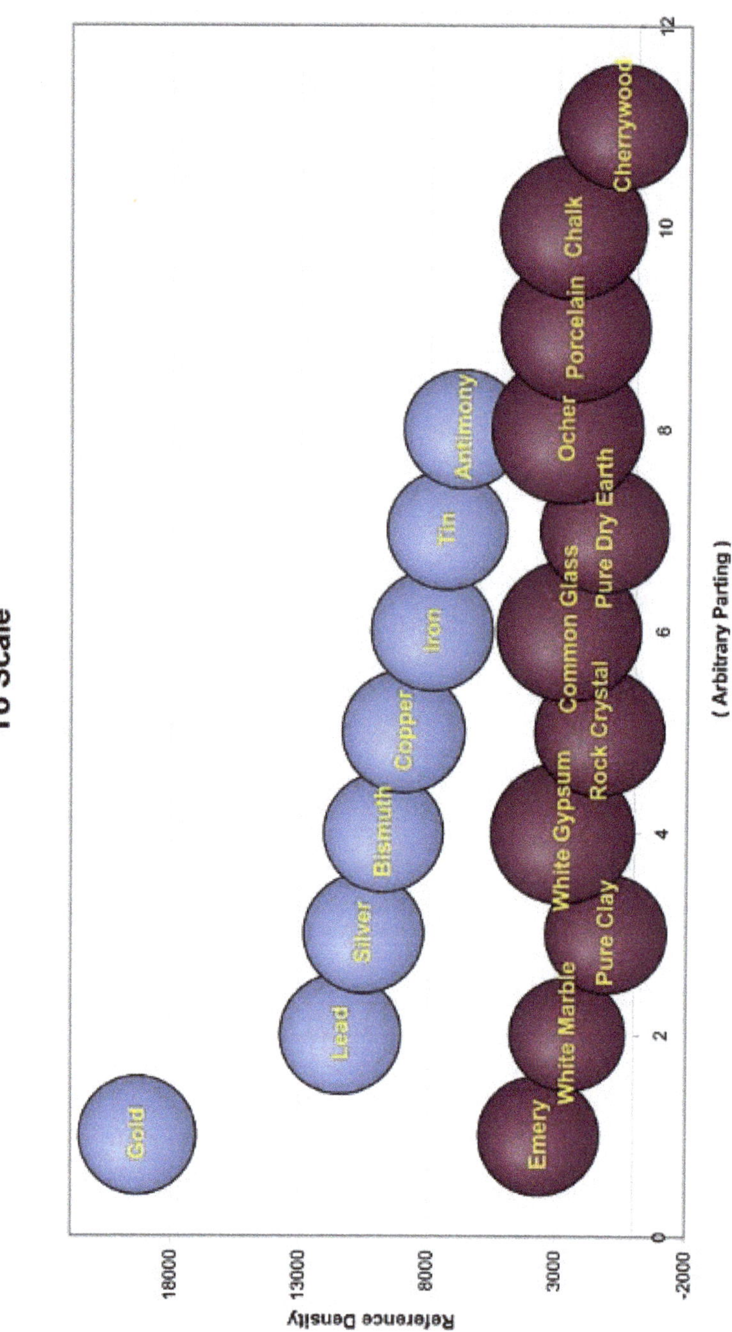

Figure Two
Tillet's Globe Sizes to Scale
As a 3D Bubble Chart

APPENDIX ONE

Facsimiles of the Barr's Buffon Text
Relative to Monsieur Tillet's Globe Fabrications

relatively to their denfity ; and I have found by experience, that this progrefs, as well in folids as fluids, is made rather by reafon of their fluidity, or in an inverted ratio of their folidity. I mean by *folidity* the quality oppofite to fluidity ; and I fay, that it is in an inverted ratio of this quality that the progrefs of heat is made in moft bodies ; and that they heat or cool fo much the fafter as they are the more fluid, and fo much the flower as they are more folid, every other circumftance being equal.

To prove that folidity, taken in this fenfe, is perfectly independent of denfity, the moft or leaft denfe matters, heat or cool more readily than other more or lefs denfe matters, for example, gold or lead, which are much more denfe than iron and copper, heat and cool much quicker ; while tin and marble, which are not fo denfe, heat and cool much fafter than iron and copper ; and there are likewife many other matters which come under the fame defcription ; fo that denfity is no ways relative to the fcale of the progrefs of heat in folid bodies.

It is likewife the fame in fluids, for I have obferved, that quickfilver, which is thirteen or fourteen times more denfe than water, neverthelefs heats and cools in lefs time than water ; and

and fpirit of wine, which is lefs denfe than water, heats and cools much quicker ; fo that generally the progrefs of heat in bodies, as well for the ingrefs as egrefs, has an affinity with their denfity, and is principally made in ratio of their fluidity, by extending the fluidity to a folid ; from hence I concluded, that we fhould know the real degree of fluidity in bodies, by heating them to the fame heat ; for their fluidity would be in a like ratio as that of the time during which they would receive and lofe this heat ; and that it would be the fame with folid bodies. They will be fo much the more folid, that is to fay, fo much the more *non fluids*, as they require more time to receive and lofe this heat, and that almoft generally to what I prefume ; for I have already tried thefe experiments on a great number of different matters, and from them I made a table, which I have endeavoured to render as complete and exact as poffible.

We caufed feveral globes to be made of an inch diameter from the following matters, which nearly reprefent the Mineral kingdom.

M. Tillet, of the Academy of Sciences, made the globe of refined gold at my particular requeft, request, and the whole of them weighed as follows :

	oz.	d.	gr.
Gold — — —	6	2	17
Lead — —	3	6	28
Pure silver —	3	3	22
Bismuth — — —	3	0	3
Copper-red — —	2	7	56
Iron — — —	2	5	10
Tin — — —	2	3	48
Antimony melted, and which had small cavities on its surface	2	1	34
Fine — — —	2	1	2
Emery — — —	1	2	24½
White marble — —	1	0	25
Pure clay — —	0	7	24
Marble common of Montbard	0	7	20
White gypsum, improperly called Alabaster — —	0	6	36
Calcareous white stone of the quarry of Anieres, near Dijon —	6	6	6
Rock chrystal; it was a little too small, and had many defects. I presume that without them it would have weighed —	0	6	22
Common glass — —	0	6	21

Pure

	oz.	d.	gr.
Pure earth, very dry — —	0	6	16
Oker — — —	0	5	9
Porcelain of the Court de Laura-guais — —	0	5	2½
White chalk — —	0	4	49
Cherrywood, which although lighter than most other woods, is that which takes in the least fire	0	1	59

I must here observe, that a positive conclusion must not be made of the exact specific weight of each matter from the preceding table, for notwithstanding the precaution that was taken to render the globes equal, as I was obliged to employ different workmen, some were too large, and others too small. Those which were more than an inch diameter were diminished, but those of rock chrystal, glass and porcelain, which were rather too small, we suffered to remain, and only rejected those of agate, jasper, and porphyry, which were sensibly so. This precision in size was not absolutely necessary, for it could very little alter the result of my experiments.

Previous to ordering these globes, I exposed to a like degree of fire, a square mass of iron, and another of lead of two inches diameter, and found, by reiterated essays, that lead heated and

APPENDIX TWO

Some Mathematical Aspects of the
Gosset-Welch Probability Distribution

I employed 64-bit MATHCAD Express$^®$ to explore the numerical integration of Gosset-Welch functional components, none of which have analytic solutions.

The various trials and experiments are embodied in my 28 May 2016 Express worksheet TYIELDERh.MCDX that is not presented here.

A User-Defined Function for Gamma

PRC MATHCAD Express does not provide an intrinsic Gamma Function gratis, and accordingly the user is obliged to program such himself as a user-defined function. The integral definition of gamma involves an infinite upper bound which is not of course numerically resolvable. By trial and error I found that an upper bound of 1500 gave some seven-figure accuracy for a small argument and therefore I established:-

$$\Gamma(t) = \int_0^\infty x^{t-1}.e^{-x}dx \approx \int_0^{1500} x^{t-1}.e^{-x}dx$$

Equation A2.1

where the following (two-sided) test values were assumed; $v = 23$, $t = 0.1$, Critical Fiducial = 1.714 (the Critical Fiducial was not applied) and Gosset-Welch Probability $\Phi(t|v)$ = 0.921210874171631.

This identity tested my user-defined $\Gamma(t)$:-

$$\Gamma(t) = \Gamma\left(\frac{7}{2}\right) = \frac{15}{8} \cdot \sqrt{\pi} = 3.32335097044784$$

Equation A2.2

Express$^®$ $\Gamma(7/2)$ computed as 3.32335088556082 which I considered sufficiently accurate for empirical purposes.

The Stirling Series Approximation of Gamma[6]

I also trialled the Stirling-type series approximation of $\Gamma(z)$ as defined by:-

$$\Gamma(z) \approx z^{z-\frac{1}{2}}.e^{-z}.\sqrt{2\pi}.\left(1 + \frac{1}{12z} + \frac{1}{288z^2} - \frac{139}{51840z^3} - \frac{571}{2488320z^4}\right)$$

Equation A2.3

This approximated $\Gamma(7/2)$ as 3.32334627870431. This is also seven-figure accuracy but slightly inferior to the user-defined integral, so I applied the latter.

<u>Gosset Two-Sided Probability $\Phi(t/v)$</u>

The Gosset Two-Sided Probability is defined as:-

$$\Phi(t|v) = 1 - \int_{-t}^{t} \frac{\Gamma\left(\frac{v+1}{2}\right)}{\sqrt{\pi v}.\,\Gamma\left(\frac{v}{2}\right)} \cdot \left(1 + \frac{t^2}{v}\right)^{-\left(\frac{v+1}{2}\right)} .\,dt$$

Equation A2.4

This expression has no analytic solution.

Notwithstanding that, by applying my user-defined Γ, and for the given v and t, Equation A2.4 elaborated to 0.921210874166985.

This is eleven-figure accuracy.

Because of the symmetry of a t distribution we may immediately halve the computational expense by using:-

$$\Phi(t|v) = 1 - 2 \int_{0}^{t} \frac{\Gamma\left(\frac{v+1}{2}\right)}{\sqrt{\pi v}.\,\Gamma\left(\frac{v}{2}\right)} \cdot \left(1 + \frac{t^2}{v}\right)^{-\left(\frac{v+1}{2}\right)} .\,dt$$

Equation A2.5

Now allow that $t_0 = 0$ and establish a second user-defined function as:-

$$I(t_0, t) = \int_{0}^{t} \left(1 + \frac{t^2}{v}\right)^{-\left(\frac{v+1}{2}\right)} .\,dt$$

Equation A2.6

Then:-

$$\Phi(t|v) = 1 - 2 \frac{\Gamma\left(\frac{v+1}{2}\right)}{\sqrt{\pi v}.\,\Gamma\left(\frac{v}{2}\right)} \int_{0}^{t} \left(1 + \frac{t^2}{v}\right)^{-\left(\frac{v+1}{2}\right)} .\,dt = 1 - 2 \frac{\Gamma\left(\frac{v+1}{2}\right)}{\sqrt{\pi v}.\,\Gamma\left(\frac{v}{2}\right)} \cdot I(t_0, t)$$

Equation A2.7

By the Duplication Rule for the Gamma Function:-

$$\Gamma(z)\Gamma\left(z + \frac{1}{2}\right) = 2^{1-2z}.\,\sqrt{\pi}.\,\Gamma(2z)$$

Equation A2.8

By substitution in Equation A2.7:-

$$\Phi(t|v) = 1 - 2\frac{2^{1-v}.\sqrt{\pi}.\Gamma(v)}{\sqrt{\pi v}.\Gamma\left(\frac{v}{2}\right)^2} \cdot I(t_0, t)$$

Equation A2.9

which simplifies to:-

$$\Phi(t|v) = 1 - \frac{2^{2-v}.\Gamma(v)}{\sqrt{v}.\Gamma\left(\frac{v}{2}\right)^2} \cdot I(t_0, t)$$

Equation A2.10

<u>The Expression of $\Phi(t|v)$ in terms of The Exponential Theorem</u>

It is possible to express the integrand of Equation A2.7 in these terms:-

$$\left(1 + \frac{t^2}{v}\right)^{-\left(\frac{v+1}{2}\right)} = \frac{1}{\left(1 + \frac{t^2}{v}\right)^{\left(\frac{v+1}{2}\right)}} = \frac{1}{\sqrt{\left(1 + \frac{t^2}{v}\right)^{(v+1)}}}$$

Equation A2.11

Allow that:-

$$n = v + 1$$

Equation A2.12

and:-

$$x = \frac{t^2}{v}$$

Equation A2.13

The Binomial Combinatorial is defined by:-

$$C(n, k) = \frac{n!}{k!\,(n-k)!}$$

Equation A2.14

I established C(n,k) as a third user-defined function within my MATHCAD Express® worksheet.

This enabled me to develop the I(t₀,t) integrand as:-

$$\frac{1}{\sqrt{(1+x)^n}} = \frac{1}{\sqrt{\sum_{k=0}^{n} C(n,k).x^k.1^{n-k}}} = \frac{1}{\sqrt{\sum_{k=0}^{n} C(n,k).x^k}}$$

Equation A2.15

Substitution and simplification then gives:-

$$\left(1+\frac{t^2}{v}\right)^{-\left(\frac{v+1}{2}\right)} = \frac{1}{\sqrt{\sum_{k=0}^{v+1} C(v+1,k).\left(\frac{t^2}{v}\right)^k}} = \frac{1}{\sqrt{\sum_{k=0}^{v+1} C(v+1,k).\left(\frac{t^{2-k}}{v}\right)}}$$

Equation A2.16

By substitution of Equation A2.10 in Equation A2.16 we may express the Probability $\Phi(t|v)$ as:-

$$\Phi(t|v) = 1 - \frac{2^{2-v}.\Gamma(v)}{\sqrt{v}.\Gamma\left(\frac{v}{2}\right)^2} \cdot \int_0^t \frac{1}{\sqrt{\sum_{k=0}^{v+1} C(v+1,k).\left(\frac{t^{2-k}}{v}\right)}}.dt$$

Equation A2.17

Equation A2.17 also yields eleven-figure accuracy. I have no information regarding its time and cost efficiencies. I did not adopt it as I did not consider that it would be as economical as the form of Equation A2.6.

<u>Aspects of Numerical Integration[6]</u>

I essayed the accuracy of several types of Newton-Cotes Closed Form integral approximating functions for Probability $\Phi(t|v)$.

These were 6, 8 and 100-interval implementations of the Extended Simpson's Rule[7]; and the 7 and 15-interval forms of A&S Rule 18[8].

All were established as further Express® user-defined functions.

The integrand was defined as:-

$$f(i, h) = \left(1 + \frac{(ih)^2}{v}\right)^{-\left(\frac{v+1}{2}\right)}$$
Equation A2.18

where for trial purposes $v = 23$ and:-

$$h = \frac{t}{n}$$
Equation A2.19

where h is the Series Interval and n is the Number of Intervals.
In all instances the Percentage Specific Defect, $\%\eta$, is defined as:-

$$\%\eta = 100\left[\frac{I_{NC}(n, t) - I(t_0, t)}{I(t_0, t)}\right]$$
Equation A2.20

The Fiducial $I(t_0, t)$ defined by Equation A2.6 for an upper bound t of 0.1 was 0.099826381425708 according to my Express® user-defined function.

All Newton-Cotes formulae are scaled summations of coefficient-weighted ordinates, taken at equal abscissal intervals. These ordinates may be empirical data, or computed integrands whether the associated integrals are analytic or not.

All these summations are finite and, at least for non-analytic integrands, all involve residual error: In general, error diminishes precipitately with the order of the series module, and slowly with the number of intervals in extension.

(A)(i) Simpson's Rule Primitive Module[7] (for n = 2)
This is defined by:-

$$IsimpPM(n, t) = \frac{h}{3}\{f(0, h) + 4f(1, h) + f(2, h)\}$$
Equation A2.21

Without extension the Simpson's module already yields a tidy seven-figure accuracy for the specified $\Phi(t|v)$.

Please consult Table A3 for details regarding the accuracy of this and the other Newton-Cotes forms tested.

(A)(ii) Extended Simpson's Rule for n = 6

When a Newton-Cotes formula is extended for further intervals an integral multiple of the module is obligatory and the interior terminal coefficients are overlapped and added in the pattern illustrated (for the Simpson case) in Table A1:-

```
i    0 1 2 3 4 5 6 7 8
     1 4 1
         1 4 1
             1 4 1
                 1 4 1
     _________________________
     1 4 2 4 2 4 2 4 1   Staggered Sum
     _________________________
```

Table A1
The Manner of Simpson Module Concatenation

Therefore for intervals n = 6 we may write:-

$$Isimpn(6, t) = \frac{h}{3}\left\{ \begin{array}{c} f(0, h) + 4[f(1, h) + f(3, h) + f[5, h]] + \\ 2[f(2, h) + f(4, h)] + f(6, h)] \end{array} \right\}$$

Equation A2.22

(A)(iii) Extended Simpson's Rule for n = 8

For purposes of handling and presentation only it is convenient to segregate coefficients in his manner:-

$$a1 = 4[f(1, h) + f(3, h) + f(5,7) + f(7, h)]$$

Equation A2.23

$$a2 = 2[f(2, h) + f(4, h) + f(6,7)]$$

Equation A2.24

By substitution the integral is then:-

$$Isimp(8, t) = \frac{h}{3}\{f(0, h) + a1 + a2 + f(8, h)\}$$

Equation A2.25

(A)(iv) Extended Simpsons Rule for n = 100
For larger numbers of intervals it is practical to employ summary summations:-

$$Isimp(100, t) = \frac{h}{3}\left\{ f(0, h) + 4\sum_{i=1}^{n/2} f(2i - 1, h) + 2\sum_{i=1}^{\frac{n}{2}-1} f(2i, h) + f(n, h) \right\}$$

Equation A2.26

(A)(v) Extended Simpson's Rule for High n
When the number of intervals becomes very large it is sometimes possible to control error by suppressing the number of individual summative events.
To this end I contrived the alternator:-

$$C_i = 3 + (-i)^{i+1}$$

Equation A2.27

and installed it thus:-

$$Isimp(n, t) = \frac{h}{3}\left\{ f(0, h) + f(n, h) + \sum_{i=1}^{n-1} [3 + (-1)^{i+1}].f(i, h) \right\}$$

Equation A2.28

The development is seen in Table A2:-

i	u (-1)^(i+1)	v 3+u
1	1	4
2	-1	2
3	1	4
4	-1	2
5	1	4
6	-1	2
7	1	4

Table A2
The Development of the 4, 2 Alternator

(B)   Primitive and Extended Rule 18

The primitive module of A&S Rule 18 is defined for eight intervals. For both this and its 16-interval extension it is convenient to partition the weighted ordinates in the following way:-

$$b1 = 989f(0,h) + 5888f(1,h) - 928f(2,h) + 10496f(3,h)$$
Equation A2.29

$$b2 = -4540f(4,h) + 10496f(5,h) - 928f(6,h) + 5888f(7,h)$$
Equation A2.30

$$b3 = 1978f(8,h) + 5888f(9,h) - 928f(10,h) + 10496f(11,h)$$
Equation A2.31

$$b4 = -4540f(12,h) + 10496f(13,h) - 928f(14,h) + 5888f(15,h)$$
Equation A2.32

The primitive Rule 18 (for n = 8) is accordingly:-

$$Ir18(8,t) = \frac{4h}{14175}\{b1 + b2 + 989f(8,h)\}$$
Equation A2.33

whilst the n = 16 Extended Rule 18 is:-

$$Ir18(16,t) = \frac{4h}{14175}\{b1 + b2 + b3 + b4 + 989f(16,h)\}$$
Equation A2.34

The n = 8 and n = 16 results for these test data were identical to the Express® fiducial integral.

The accuracy of these various methods is displayed in Table A3:-

Newton-Cotes Modular Form	A&S Number	ν	n	Fiducial Integral	NC Integral	Percentage Specific Defect	Notes
Simpson's Rule	25.4.6	23	2	0.099826381425708	0.099826393610931	0.000012206415602	Primitive
Simpson's Rule	25.4.6	23	6	0.099826381425708	0.099826381575901	0.000000150454213	Extended
Simpson's Rule	25.4.6	23	8	0.099826381425708	0.099826381473226	0.000000047600647	Extended
Simpson's Rule	25.4.6	23	100	0.099826381425708	0.099826381425710	0.000000000002002	Extended
Simpson's Rule	25.4.6	23	8	0.099826381425708	0.099826381473226	0.000000047600647	Alternating Form
Simpson's Rule	25.4.6	23	1000	0.099826381425708	0.099826381425708	0.000000000000014	Alternating Form
Rule 18	25.4.18	23	8	0.099826381425708	0.099826381425708	-0.000000000000070	Primitive
Rule 18	25.4.18	23	16	0.099826381425708	0.099826381425708	0.000000000000028	Extended

Express and EXCEL Specific Defects sometimes Differ
in the Thirteenth, Fourteenth and Fifteenth Decimal Places
For Simpson's Rule, EXCEL defects are shown,
except for Simpson's Rule n=1000

Table A3
Comparative Accuracy of the Tested
Newton-Cotes Integrations

APPENDIX THREE

Welch's Unequal Variances
Worksheet Tabulation Equations

	A	B
1	WELCH'S UNEQUAL VARI.	
2	PHASE TWO: GROUP B C	
3		
4	TAILS (1 or 2):	2
5		
6	Computed Values:	
7	t	=ABS((D27-E27)/SQRT(D29+E29))
8	ν (approx)	=(D29+E29)^2/(D29^2/D26+E29^2/E26)
9	p=A(t|ν) (TDIST)	=TDIST(B7,B8,B4)
10	t (TINV)	=TINV(B9,B8)
11		
12	Mean of Differences, μ_δ	=AVERAGE(G34:G59)
13	Variability, σ_{rel}	=IF(D25<>E25,"UNDEFINED",STDEVP(G34:G59)/ABS(B12))
14	Precision, Ω	=IF(D25<>E25,"UNDEFINED",1/(1+B13))
15		
16	Group 1	Tillet's Series
17	Group 2	Fiducial Series
18		
19		
20		
21		
22		
23		
24		
25		
26		
27		
28		
29		
30		
31		
32		
33		
34		Emery
35		White Marble
36		Pure Clay
37		White Gypsum
38		Rock Crystal (defective)
39		Common Glass
40		Pure Dry Earth
41		Ocher
42		Lauraguais Porcelain
43		White Chalk
44		Cherrywood
45		
46		
47		
48		
49		
50		
51		
52		
53		
54		
55		
56		
57		
58		
59		

	C	D	E	F
1				
2				
3				
4				
5				
6				
7				
8	Welch-Satterthwaite Equation			
9				
10				
11				
12				
13	$\sigma_{rel} = (\Sigma(\delta-\mu_\delta)^2/N_2)^{0.5}/|\mu|$			
14	$\Omega = 1/(1+\sigma_{rel})$			
15				
16				
17				
18				
19		Group 1	Group 2	
20		FALSE	TRUE	
21				
22	μ_1	=AVERAGE(D34:D59)	=AVERAGE(E34:E59)	μ_2
23	σ_1	=STDEVP(D34:D59)	=STDEVP(E34:E59)	σ_2
24				
25	N_1	=COUNT(D34:D59)	=COUNT(E34:E59)	N_2
26	ν_1	=D25-1	=E25-1	ν_2
27	$Xbar_1$	=AVERAGE(D34:D59)	=AVERAGE(E34:E59)	$Xbar_2$
28	s_1^2	=VAR(D34:D59)	=VAR(E34:E59)	s_2^2
29	s_1^2/N_1	=D28/D25	=E28/E25	s_2^2/N_2
30				
31	Serial	Group 1	Group 2	
32		FALSE	TRUE	
33				
34	0	3317.54837756223	3650	
35	=C34+1	3074.55138768682	2550	
36	=C35+1	982.219411706937	1089	
37	=C36+1	920.830698475253	2787	
38	=C37+1	849.210533038289	1390	
39	=C38+1	844.094806935649	2579	
40	=C39+1	818.516176422448	1249	
41	=C40+1	659.928667240598	2700	
42	=C41+1	626.676447573436	2403	
43	=C42+1	741.780284882843	2499	
44	=C43+1	424.605266519145	630	
45	=C44+1			
46	=C45+1			
47	=C46+1			
48	=C47+1			
49	=C48+1			
50	=C49+1			
51	=C50+1			
52	=C51+1			
53	=C52+1			
54	=C53+1			
55	=C54+1			
56	=C55+1			
57	=C56+1			
58	=C57+1			
59	=C58+1			

	G	H
1		
2		
3		
4		
5		
6		
7		
8		
9		
10		
11		
12		
13		
14		
15		
16		
17		
18		
19		
20		
21		
22		
23		
24		
25		
26		
27		
28		
29		
30		
31	Difference, δ	$(\delta - \mu_\delta)^2$
32		
33		
34	=IF(OR(ISBLANK(D34),ISBLANK(E34)),"",E34-D34)	=IF(OR(ISBLANK(D34),ISBLANK(E34)),"",G34^2)
35	=IF(OR(ISBLANK(D35),ISBLANK(E35)),"",E35-D35)	=IF(OR(ISBLANK(D35),ISBLANK(E35)),"",G35^2)
36	=IF(OR(ISBLANK(D36),ISBLANK(E36)),"",E36-D36)	=IF(OR(ISBLANK(D36),ISBLANK(E36)),"",G36^2)
37	=IF(OR(ISBLANK(D37),ISBLANK(E37)),"",E37-D37)	=IF(OR(ISBLANK(D37),ISBLANK(E37)),"",G37^2)
38	=IF(OR(ISBLANK(D38),ISBLANK(E38)),"",E38-D38)	=IF(OR(ISBLANK(D38),ISBLANK(E38)),"",G38^2)
39	=IF(OR(ISBLANK(D39),ISBLANK(E39)),"",E39-D39)	=IF(OR(ISBLANK(D39),ISBLANK(E39)),"",G39^2)
40	=IF(OR(ISBLANK(D40),ISBLANK(E40)),"",E40-D40)	=IF(OR(ISBLANK(D40),ISBLANK(E40)),"",G40^2)
41	=IF(OR(ISBLANK(D41),ISBLANK(E41)),"",E41-D41)	=IF(OR(ISBLANK(D41),ISBLANK(E41)),"",G41^2)
42	=IF(OR(ISBLANK(D42),ISBLANK(E42)),"",E42-D42)	=IF(OR(ISBLANK(D42),ISBLANK(E42)),"",G42^2)
43	=IF(OR(ISBLANK(D43),ISBLANK(E43)),"",E43-D43)	=IF(OR(ISBLANK(D43),ISBLANK(E43)),"",G43^2)
44	=IF(OR(ISBLANK(D44),ISBLANK(E44)),"",E44-D44)	=IF(OR(ISBLANK(D44),ISBLANK(E44)),"",G44^2)
45	=IF(OR(ISBLANK(D45),ISBLANK(E45)),"",E45-D45)	=IF(OR(ISBLANK(D45),ISBLANK(E45)),"",G45^2)
46	=IF(OR(ISBLANK(D46),ISBLANK(E46)),"",E46-D46)	=IF(OR(ISBLANK(D46),ISBLANK(E46)),"",G46^2)
47	=IF(OR(ISBLANK(D47),ISBLANK(E47)),"",E47-D47)	=IF(OR(ISBLANK(D47),ISBLANK(E47)),"",G47^2)
48	=IF(OR(ISBLANK(D48),ISBLANK(E48)),"",E48-D48)	=IF(OR(ISBLANK(D48),ISBLANK(E48)),"",G48^2)
49	=IF(OR(ISBLANK(D49),ISBLANK(E49)),"",E49-D49)	=IF(OR(ISBLANK(D49),ISBLANK(E49)),"",G49^2)
50	=IF(OR(ISBLANK(D50),ISBLANK(E50)),"",E50-D50)	=IF(OR(ISBLANK(D50),ISBLANK(E50)),"",G50^2)
51	=IF(OR(ISBLANK(D51),ISBLANK(E51)),"",E51-D51)	=IF(OR(ISBLANK(D51),ISBLANK(E51)),"",G51^2)
52	=IF(OR(ISBLANK(D52),ISBLANK(E52)),"",E52-D52)	=IF(OR(ISBLANK(D52),ISBLANK(E52)),"",G52^2)
53	=IF(OR(ISBLANK(D53),ISBLANK(E53)),"",E53-D53)	=IF(OR(ISBLANK(D53),ISBLANK(E53)),"",G53^2)
54	=IF(OR(ISBLANK(D54),ISBLANK(E54)),"",E54-D54)	=IF(OR(ISBLANK(D54),ISBLANK(E54)),"",G54^2)
55	=IF(OR(ISBLANK(D55),ISBLANK(E55)),"",E55-D55)	=IF(OR(ISBLANK(D55),ISBLANK(E55)),"",G55^2)
56	=IF(OR(ISBLANK(D56),ISBLANK(E56)),"",E56-D56)	=IF(OR(ISBLANK(D56),ISBLANK(E56)),"",G56^2)
57	=IF(OR(ISBLANK(D57),ISBLANK(E57)),"",E57-D57)	=IF(OR(ISBLANK(D57),ISBLANK(E57)),"",G57^2)
58	=IF(OR(ISBLANK(D58),ISBLANK(E58)),"",E58-D58)	=IF(OR(ISBLANK(D58),ISBLANK(E58)),"",G58^2)
59	=IF(OR(ISBLANK(D59),ISBLANK(E59)),"",E59-D59)	=IF(OR(ISBLANK(D59),ISBLANK(E59)),"",G59^2)

APPENDIX FOUR

Tillet's Globes Size Properties

Table A4.1
Tillet's Globes Size Properties

Unit Mass (kgs)　　3.E-02　1.E-03　5.E-05

Material	oz.	d.	gns.	Globe Mass (kilograms)	Computed Tillet Density (kg/m^3)	Reference Density (kg/m^3)	Percentage Specific Defect	Notes	Group	Globe Volume (m^3)	Globe Radius (m)	Globe Diameter (m)
Gold	6	2	17	0.18701492	18012.47161	19320	-6.767745303		A	9.67986E-06	0.013221	0.026442
Lead	3	6	28	0.10091688	9719.879595	11389	-14.65554838		A	8.86091E-06	0.012837	0.025674
Silver	3	3	22	0.09677398	9320.852959	10500	-11.22997182		A	9.21657E-06	0.013006	0.026013
Bismuth	3	0	3	0.09194059	8855.321884	9747	-9.148231418		A	9.43271E-06	0.013107	0.026215
Copper	2	7	56	0.07308507	7039.239117	8960	-21.43706342		A	8.15682E-06	0.012488	0.024975
Iron	2	5	10	0.06809234	6558.360864	7874	-16.70865045		A	8.64774E-06	0.012733	0.025466
Tin	2	3	48	0.06756120	6507.203603	7310	-10.98216686	(α) 5750 (β) 7310	A	9.2423E-06	0.013019	0.026037
Antimony	2	1	34	0.06426812	6190.028584	6691	-7.487242801		A	9.60516E-06	0.013187	0.026374
Emery	1	2	24.5	0.03444453	3317.548378	3650	-9.108263628	corundum	B	9.43686E-06	0.013109	0.026219
White Marble	1	0	25	0.03192160	3074.551388	2550	20.57064265		B	1.25183E-05	0.014404	0.028808
Pure Clay	0	7	24	0.01019792	982.2194117	1089	-9.805380009		B	9.36448E-06	0.013076	0.026151
White Gypsum	0	6	36	0.00956055	920.8306985	2787	-66.95978836		B	3.43041E-06	0.009356	0.018712
Rock Crystal	0	6	22	0.00881695	849.210533	1390	-38.90571705	flint	B	6.34313E-06	0.011483	0.022967
Common Glass	0	6	21	0.00876383	844.0948069	2579	-67.27046115		B	3.39815E-06	0.009326	0.018653
Pure Dry Earth	0	6	16	0.00849826	818.5161764	1249	-34.46627891	dry loam	B	6.80405E-06	0.011755	0.02351
Ocher	0	5	9	0.00685173	659.9286672	2700	-75.55819751	limonite: sg=2.7-4.3	B	2.53768E-06	0.008462	0.016923
Porcelain	0	5	2.5	0.00650648	626.6764476	2403	-73.92108		B	2.70765E-06	0.008646	0.017293
Chalk	0	4	49	0.00770155	741.7802849	2499	-70.31691537		B	3.08185E-06	0.009028	0.018055
Cherrywood	0	1	59	0.00440847	424.6052665	630	-32.60233865		B	6.99758E-06	0.011866	0.023731

Mean Percentage Specfic Defect (All)　　　　-29.30317886
Mean Percentage Specfic Defect (Metals)　　-12.30207756
Mean Percentage Specfic Defect (Non-Metals)　　-41.66761618

www.ingramcontent.com/pod-product-compliance
Lightning Source LLC
Chambersburg PA
CBHW041110090726
47602CB00020B/50